Nick E. Flynn
Chemist Brewers

Also of interest

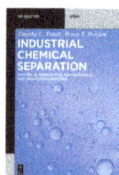

Industrial Chemical Separation.
Historical Perspective, Fundamentals, and Engineering Practice
Timothy C. Frank and Bruce S. Holden, 2023
ISBN 978-3-11-069502-1, e-ISBN (PDF) 978-3-11-069505-2

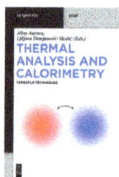

Thermal Analysis and Calorimetry.
Versatile Techniques
Edited by: Aline Auroux and Ljiljana Damjanović-Vasilić, 2023
ISBN 978-3-11-059043-2, e-ISBN (PDF) 978-3-11-059044-9

Chemical Technicians.
Good Laboratory Practice and Laboratory Information Management Systems
Mohamed Elzagheid, 2023
ISBN 978-3-11-119110-2, e-ISBN (PDF) 978-3-11-119149-2

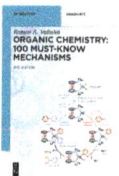

Organic Chemistry: 100 Must-Know Mechanisms
Roman Valiulin, 2023
ISBN 978-3-11-078682-8, e-ISBN (PDF) 978-3-11-078683-5

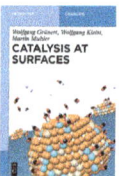

Catalysis at Surfaces
Wolfgang Grünert, Wolfgang Kleist and Martin Muhler, 2023
ISBN 978-3-11-063247-7, e-ISBN (PDF) 978-3-11-063248-4

Nick E. Flynn

Chemist Brewers

Insights from Chemists and Biologists in the Brewing
Industry

DE GRUYTER

Author
Prof. Dr. Nick Edward Flynn
Department of Chemistry/Physics
West Texas A&M University
NSB 114E, WTAMU Box
Canyon, TX 79016
USA
nflynn@wtamu.edu

ISBN 978-3-11-079875-3
e-ISBN (PDF) 978-3-11-079877-7
e-ISBN (EPUB) 978-3-11-079894-4

Library of Congress Control Number: 2023950662

Bibliographic information published by the Deutsche Nationalbibliothek
The Deutsche Nationalbibliothek lists this publication in the Deutsche Nationalbibliografie; detailed
bibliographic data are available on the Internet at http://dnb.dnb.de.

Preface

Out of all the research and publications I have done throughout my career, this one has got to be the most fun. I had the opportunity to learn about brewing cultures and environments in several areas of the United States. The idea for the book came from a fermentation sciences course that I developed at WT. Several of our majors have asked me about brewing and how they can get into it on a professional level. As a result, I thought it would be great to interview chemists and biologists to help people learn from their experiences in the brewing industry. We start off with chemists, transition into biochemists and then go to biologists. We finish with perspectives from other disciplines to round out the interviews.

Here is a list of people who need to read this particular book:
– Career counselors
– STEM majors interested in the brewing industry
– STEM graduates interested in entering the brewing industry
– Homebrewers
– Brewery managers and owners

Each of the interviews will cover the areas below though some interviews branch off into additional areas (e.g., hops). Each interview is unique, though, and questions vary in some areas, based on the flow of the interview:
– College/degree pursuit
– Movement to brewing field
– Current position
– Experiences in the brewing industry
– Professional affiliations
– Personal interest (other/miscellaneous)

Another very special feature of this book is the QR code and website links to provide additional information and increase the level of interactivity of the book. I sincerely hope you enjoy reading this book and hope you learn more about the brewing industry from it.

Prost!
Nick Flynn
Professor of Biochemistry
West Texas A&M University

https://doi.org/10.1515/9783110798777-202

Acknowledgments

I would like to thank my wife, Serena, as well as my sons, Logan and Dillon, for being such a wonderful part of my life. Everything I do, I do for you all. I would also like to thank my parents for supporting me in getting a college education. Their support, encouragement and guidance were so important to me graduating and becoming the person who I am today.

https://doi.org/10.1515/9783110798777-203

Contents

Section II: **The biochemists**

Section III: **The biologists**

Section IV: **Other disciplines**

Section I: **The chemists**

1 Phillip (Phil) Chou

BS Chemistry, University of Utah (1984)

MS Chemistry, University of Wyoming (1986)

PhD Chemistry, University of Minnesota (1992)

John I. Haas, Inc., Yakima, WA, Director of Brewing Solutions

John Haas, Inc https:// www.johnihaas.com/ 08/28/23	University of Utah https://www.utah. edu/ 08/28/23	University of Wyoming https://www.uwyo.edu/ 08/28/23	University of Minnesota https://twin-cities.umn.edu/ 08/28/23

1.1 College/degree pursuit

1.1.1 What inspired you to pursue a chemistry (STEM) degree?

I always had an interest in science even as a kid. So it came down to a choice between biology, chemistry and physics. I chose chemistry because it is everywhere in your life and most relatable to me at least. When I compared it with biology I thought the job prospects in chemistry were much better in chemistry than in biology. I didn't think I was smart enough to do physics at the time.

1.1.2 How long did it take you to graduate with your BS degree? Were there any stumbling blocks you'd like to share?

It took me 5 years. Not really. It was just a matter of finding what part of chemistry I liked the best. For me, I realized that I like organic chemistry as an undergrad. It was more of a challenge to figure out where my interests were.

1.1.3 Did you do any research as an undergraduate? If so, what was the project?

I did a little bit. It was more like doing pharmaceutical research. So I did some lab work up at the pharmacy school and also did some lab work at a start-up that was

https://doi.org/10.1515/9783110798777-001

part of the research park at the University of Utah. We were looking at sources of more natural pesticides at that particular position.

Natural pesticides

Yep, this is exactly what it sounds like. Using what can be found in nature to repel or kill insects. A number of different plants can be used to provide some of the pesticides, including chrysanthemums, garlic and peppers.

https://www.webmd.com/a-to-z-guides/what-to-know-about-natural-pesticides

08/28/23

1.1.4 Could you talk about the work you did for your master's and PhD degrees?

I'm a physical organic chemist and that was really where I wanted to be. I was not interested in being a synthetic organic chemist making new drugs or anything like that. Mechanisms, structure and activity relationships were the things that interested me the most. When I went to Wyoming I got involved in a project related to surfactant chemistry, and the end goal was to use these vesicles as drug delivery devices. My project there was synthesizing new compounds that you can use in vesicles. My part of the project was to synthesize the building blocks of the vesicles. Then the design was such that when you saponify the vesicles, the resulting constituents would not form a difficult emulsion that was hard to work with. That was the essence of my master's work.

When I went to Minnesota for my PhD, I ventured off into the more esoteric side of chemistry. I studied gas-phase ion molecule chemistry. We looked at kinetics of reactions, reaction mechanisms in the gas phase using a home-built instrument called a flowing afterglow where you used a flow tube and generated ions in the gas phase. Flowing afterglow consists of a stainless steel tube with an ion source at the beginning, and various inlets for reagent addition, coupled to a quadrupole mass spectrometer. The system is under vacuum. Ion molecule chemistry occurs within the tube as the initially generated ions flow down the tube carried by helium gas, and product ions are detected by the mass spectrometer. That was fascinating and very different from the traditional organic chemistry that I had done before. It was nice though because you could do a lot of reactions during the day and you didn't have to do column chromatography or other complicated things like that. Ahead of that there was also a synthetic portion where you would have to synthesize the molecules that you put into the flowing afterglow. So, it was that synthetic component where I got to synthesize some interesting things like bicyclobutane and cyclopropane with different constituents on it. There was that synthetic portion to it. The goal was to study the reactivity and the chemistry and not focus so much on the synthetic end.

Gas-phase ion thermochemistry
This website provides quite a bit of information regarding gas-phase ion
thermochemistry. It's worth a view if you want to learn more about this fascinating field.
https://webbook.nist.gov/chemistry/ion/
08/28/23

1.1.5 Do you have any words of wisdom for students wishing to pursue a chemistry (STEM) degree?

My advice to them would be to not really worry about what would get them the best
job. Rather they should do what they are interested in. Most people, at least in industry, do not do what they did as a student in grad school. Find a project that is most
interesting to you, do well on that project and it will teach you how to think and how
to solve problems. Don't worry about if I do this then I'll get this job. You'll get the job
at the end of the day.

1.1.6 How often do you use what you learned as an undergraduate in your current job?

As an undergraduate? Probably a little bit. A lot of what I do now in my area is organic chemistry. Analytical chemistry is also a part of what we do. Those are probably
the two things that apply to my current position.

1.2 Movement to brewing field

1.2.1 What was your "entry beer" into enjoying craft beer? Why did you like it?

I would say that back in the day when I first started there wasn't much around. One
of the first craft beers I remember was Boulder beer. It is no longer around actually.
It was interesting because it had a lot of flavor. It was bottle fermented so there was
yeast at the bottom. So, if you were drinking straight from the bottle you had to be
careful not to get the sediment. You don't want to get all the yeast in one big slug.
Throughout my beer drinking life, I gravitated toward the craft beers that to me were
more flavorful. I was not a big "pound the commercial lager" kind of guy. What I
liked about Boulder was it did have flavor and a hoppy taste to it. Also, like I said
there weren't many around and there weren't too many craft beers that you could get
then. That drove those purchasing decisions.

Boulder beer
It is true that the company that Dr. Chou is familiar with is no longer in existence. The History tab on this site helps explain how the company was started by two professors, the closing of their original brewpub and how Boulder Social now brews some of those beers that Dr. Chou likely enjoyed.
https://www.boulderbeer.com/beersofthepast
08/28/23

1.2.2 Did you get started with homebrewing?

Not really. When I first got into beer, homebrewing was not really a thing. This was in the early 1980s. So I would enjoy beer more as a consumer than a producer back then.

1.2.3 Prior to getting into brewing what else did you do after graduating?

My initial goal coming out of graduate school was to get an academic job. I did a couple of postdoctorates. I did one at the University of Pittsburgh and that was more traditional physical organic chemistry looking at biradical chemistry. We would synthesize precursors. I did a second postdoctorate and that was more in the gas-phase chemistry area. I did that at Purdue and so I then changed my mind and wanted to go into industry. My first job was working for Great Lakes Chemical in West Lafayette, Indiana. The projects I was working on focused on developing new flame-retardant additives in plastics and foam. This again points to the stuff that I did in graduate school for the most part did not really relate to developing flame retardants. That was an interesting job. I learned a lot about it. You do not really think about it, but things like TVs and car interiors are basically solidified fuel waiting to go up. I had a lot of fun and learned a lot of things. I got to catch many things on fire and also got exposed to things like ASTM methods and things like that. You know the things that you would normally not be exposed to as an academic chemist.

Great Lakes Chemical
Talk about your corporation treasure hunt. Great Lakes was a standalone company until it was merged with Chemtura in 2005. Then in 2017 Chemtura was acquired by Lanxess.
https://lanxess.com/en/Products-and-Brands/Brands/Reolube
08/28/23

I did that for about 4 years and then my entry into brewing was not planned or intentional. I saw an advertisement in *C&E News* for an R&D (research and development) chemist position at Miller Brewing Company. I was certainly not looking to leave at that point, but I thought, "You know, I like beer. I probably won't get the job, but I'll apply for it." I was surprised when I got it. This reinforces that you should do what you're interested in because you never know where you are going to end up. If you had told me during grad school or during my first job that I'm going to be working in beer, I would not think that was how it would happen. That is, really how I got started in beer was more so by happenstance than anything else.

1.2.4 Would you recommend brewing or the hops industry to other chemists out there and why?

For sure. Brewing is about chemistry to a large degree. You have flavor chemistry, aroma chemistry and biochemistry going on. You have a whole range of compound classes associated with beer. You have hop bittering acids all the way to polyphenols to terpenes to sesquiterpenes. There is still quite a bit that we do not know a lot about in regard to beer chemistry. Many opportunities, especially from an analytical chemistry point of view. There are certainly many areas of chemistry associated with beer.

1.3 Current position

1.3.1 What are your current job responsibilities and tasks?

My title is director of Brewing Solutions. When I first got to Haas, my original position was director of R&D. I worked on developing new hop-based products, but then Haas created this new team called Brewing Solutions and that seemed really intriguing to me. My whole career had been R&D before Haas. There is a whole host of different activities that go on with my current team. There is a sensory program that we do. Everything that we produced from new product development, to extracts, to experimental strains they have to be brewed with and then run through sensory. The proof is in the pudding right? You have to make sure that the flavors and the aromas that you are going to get that you are getting. Sensory is a big part of our group because of that. We also have a nanobrewery. It is a 20 gallon brewery that is basically an advanced homebrew setup. There we do a lot of testing of hop brewing with new varieties that come from our brewing program. We look at new products that come out of our R&D group. We also do what we call competitive intelligence. All of those things end up getting brewed on that system and it goes through sensory. There is also an educational component that goes through us, where we offer a variety of classes. One is called Hops Academy, where it is hops A–Z over a couple of days that we offer to the industry. Then we have a more sensory-focused class called hops flavorist, where we take a sensory training program down to 3 days. Another piece that we do is the quality end of things. Our team is responsible for doing analysis on every lot that comes in. I think we did 1,200 or 1,300 different lots this last season. Then we do technical support and answer customer questions about hops. We do training on new and existing products too, so we wear many hats in the group. The thing that appeals to me is it gets me out of that R&D role that I've been doing most of my life. It is more customer-facing and it is the fun part of this for me. I still have a foot in the technical side of things. I did not want to go into something like sales though.

John Haas
This company generates hops and hop products for brewers. They are affiliated with several organizations including Oregon State University, American Society of Brewing Chemists (ASBC) and the USA Hops.
https://www.johnihaas.com/
08/28/23

1.3.2 What classes that you took apply to some of the things that you have done in the brewing industry? Why/how?

Analytical chemistry of course. We use GC and HPLC and even down to the more mundane things like measuring pH and using a UV–Vis spectrophotometer. It is important for us to characterize our hops and hop products. The hop bittering acids and things like that such as polyphenols. Those are huge from the education standpoint. Organic chemistry, of course, plays a big role. The important compounds we deal with from a bittering, aroma and flavor standpoint are all organic compounds. Then physical chemistry even plays a role in trying to understand from an extraction standpoint what would be the best solvent standpoint. We also do supercritical extractions of hops that involves a physical chemistry component as well. The one that I probably do not use much of in this particular position is inorganic chemistry.

1.3.3 What adjustments did you need to make from working in an academic lab setting to the hops industry?

The big adjustment in moving from an academic setting to industry is that you are not just doing things for knowledge. You need to contribute to the bottom line of the company. So, the sales and business component is something we do not get much training on in academia. Budgeting becomes an important thing that we were not trained on. Sometimes the sense of urgency is very different in industry compared to academia. I have encountered people coming just out of school and they have not seen that before. You cannot just sit on a problem or a project and you have to go at a pace that you may not be used to. Time management probably becomes more important in that aspect.

1.3.4 What do you enjoy most about your current position?

Being able to work on a product that is fun to make and that people enjoy. You can go to the store and point to things that you played a part in developing or making. Commercializing a beer that had products that I helped work on is enjoyable. People want to talk to you about it. Say you're at a party and people ask you what you do. If you told them flame retardants, they would probably walk away. Now, if you tell them that you work on beer, everybody wants to talk to you about beer.

1.3.5 Do you have any advice for students wishing to apply their chemistry (STEM) degree toward getting into the hops industry?

If that is something that you are interested in keep your eyes out for them and apply. We're currently looking for a PhD chemist and will eventually be looking for a BS chemist. So, the hops industry is pretty small. There are not tons of openings, but they are there. Look at different resources such as the ASBC job board or Master Brewers Association of the Americas (MBAA) or LinkedIn. Target your search and keep your eyes open. Maybe get a job at a brewery and then it might be easier to move on to hops. Really you do not even need to work in the brewing industry. You can work as a chemist in another industry outside of beer or hops. Do some real-world analytical work and just keep your eyes open and apply because jobs will come up.

1.4 Hops industry

1.4.1 What drew you to the hops industry?

Like I said. It was a chemistry job and it involved beer. Actually the projects I worked on at Miller were hop projects. So, that was my first exposure to the hops industry. One of the first projects I worked on was quality of hops bitterness. You have all these hop acids and do they have a different quality of bitterness. One of the projects I worked on was the different hop compounds like humulone and cohumulone. Converting them into their different iso-alpha-derivatives or forms. Then we did sensory analysis on them. As an aside, brewers believe that hops with more cohumulone do not brew as good of a bitterness as those hops with less cohumulone. We found that was not true. What we found was that is, in reality, isocohumulone does not do better than cohumulone. In fact, some people prefer the cohumulone in hops. That was fascinating to me and that was my start.

Then I moved on to the wine industry in R&D. I looked at color, aroma and mouthfeel and how that actually developed. We looked at the vineyards and the processing of grapes. Then I decided that I wanted to get into the brewing industry so I started a small microbrewery. We then did that for 3 years then we sold it. That taught me a lot about hops, recipe development and things that helped a lot in my current role. You know, what is important from a holistic and business sense. From there I worked briefly for a food and beverage ingredient supplier and then this opportunity came along and brought me full circle into the brewing industry. It was a natural progression from getting back into hops.

1.4.2 How has the hops industry changed since you first started in it?

It has actually changed a lot. When I first got started in the field, alpha was the major consideration – from what people were brewing and what they were interested in buying. As the craft brewing boom hit, it started to shift to aroma and flavor hop components. When I first started, Galena was a big hop and it is not really anymore. Citra was not even around. I really liked Cascade even back then. Now, at least in the USA the hop market has really changed from the majority alpha-acid varieties to the aroma varieties like Citra and Mosaic. The other interesting thing is we would come out for Miller and come out to the Pacific Northwest for harvest season and you would never see a craft brewer. Just the big companies were there. Now, you have tons of craft brewers coming out for harvest. So, that has been a huge change. The craft brewing industry has really driven this change. They are constantly looking for new flavors and aromas and you know alpha is not very attractive anymore. So that has been a huge change compared to what it used to be like.

1.4.3 What is the biggest challenge to the hops industry right now?

I think the biggest challenge is new and unique flavors. That is all in the breeding process and it is slow. It takes a while for that to come up. You don't want to just breed another Citra and you want to anticipate flavors from a breeding standpoint. A second challenge is delivering flavor consistently. One of the things I did not mention previously is that early on if you mentioned hop extracts to craft brewers, they would just slam the door in your face. They felt like artists and would not even consider using them. Now, those are things that people are considering to use or actually using. Dry hopping rates that people are using such as New England IPAs. They are losing quite a bit to the vegetative matter that absorbs some of the other beer flavor. So, now the challenge is to deliver hop products that have the flavor people want that provides them with efficiency and beer loss. It improves the economics of the beer. With the competition and the economic climate they are having to cut costs and put their big boy pants on and make tough decisions. They have to consider doing things the older way, but at what cost. Those are the challenges that we are facing right now.

1.4.4 Heard of some new bittering compounds associated with dry hopping. Any comments or thoughts on that?

I do not know if it is so much new compounds. It is just a realization that other compounds such as isohumulone that are imparting bitterness. Also, the oxidized alpha- and beta-acids are contributing some bitterness – understanding the bittering impact. A lot of these compounds are offering less impact to bitterness. If you consider isohumulone as a 1, then other bittering chemicals can be higher or lower in relative sensory bitterness such as rho-isohumulone at 0.75. Other hop chemicals can be higher or lower in relative sensory bitterness. You need to understand the bitterness impact of these other compounds and their impact on the beer and how it will impact your International Bitterness Unit (IBU) measurement. Actually, I do not consider that to be such a good measure. For example, you see people calculate over 100 IBUs and that is physically impossible from a solubility standpoint. So, a lot of these compounds that are not even bitter will absorb in that UV range so your measurement and calculation are not accurate. You need to understand how that will impact your stated IBUs. That in and of itself is fascinating chemistry. Brewers are becoming more sophisticated and aware and so that is why you are hearing more talk about it.

1.4.5 Can you describe your most interesting contribution or the one you are most proud of to the hops industry?

Early on, we were not able to publish it, but it was confirmed by Tom Shellhammer at Oregon State and others showing that cohumulone is not the bad player that everyone thought it was; working to put new and innovative products out there that give you bitterness and efficiency on the hops side. Also, work that we did on products that are out on the market like Incognito that is out that you add to the whirlpool and that allows you to replace some of your dry hopping bill. The work that we are doing on the cold side that is not commercialized yet that can be impactful. One of the challenges on the cold side is trying to get oils into beer. They just are not as soluble. So we commercialized a product Spectrum that goes toward that. We have others in the pipeline that don't require special treatment to get them into solution such as ethanol or propylene glycol. Some other products that contribute impactful flavor are 100% hop derived as opposed to say water with no solvents necessary. We are trying to be more natural for lack of a better word. We are also working on sustainability products that I take a lot of satisfaction from.

Spectrum
A couple of major issues with dry hopping are loss of beer and hop creep – increase in ABV beyond modeled ABV. Spectrum is a liquid product used for dry hopping that avoids some of the common dry hopping issues.
https://www.johnihaas.com/spectrum/
08/28/23

1.5 Professional affiliations

1.5.1 What current professional organizations are you affiliated with? Why did you join them?

The American Chemical Society of course. I've been a member forever for professional development and journal access. I am also a member of ASBC, MBAA for the same reason as well as conference access and networking. They are also good for posting jobs. I am also a member of the Institute for Brewing and Distilling (IBD) and I initially joined for the certification in brewing. That is actually another thing I would recommend to those who want to get into brewing is to get some type of certification. IBD, of course, is a great place to start. I am also a member of the Institute of Food Technology (IFT), so we are not only interested in beer, but we are also interested in other products that use hops. That gives us a broader peek into what is going on the industry.

Institute for Brewing and Distilling (IBD)
The IBD has existed since 1896 and was originally called the Laboratory Club. They offer a variety of courses and qualifications, including Beer Sensory, Essentials in Distilling and Beer Recipe Development.
https://www.ibd.org.uk/
08/28/23

1.5.2 What do you enjoy most about going to their events?

I enjoy the technical talks. You know, what is new and what people are interested in. I enjoy talking to people to see what trends are out there. I also enjoy getting feedback on our products that are out there. Brewers Association is also good – I don't have a membership but Haas does. ASBC and IFT are good for the trade shows as well, so you can see what new equipment is out there.

1.5.3 How important is it to be associated with other professional organizations?

I think it is very important. It helps you network and there is more opportunity to get to know people. Also, there are job fairs associated with them so you can see what is also out there. Networking can also help you identify that as well as what you need to do to get a particular job that is out there. Just to keep up on the latest technical developments as well.

1.6 Other/miscellaneous

1.6.1 Outside of brewing what other hobbies or interests do you have?

I like to ski and backpack or hike. I am also a huge sports fan. So, I like to go to baseball, hockey and things like that.

1.6.2 Is there anything else that you'd like to share about your personal or professional life that you think people might be interested in knowing?

The message I would like to get out is do not be afraid to change, because you never know what you might stumble into. Like I said, I had no intention with my first job to go into brewing. It just kind of happened and then the other stops along the way. I learned about wine. There are a lot of similarities between wine and beer. The big change for me was opening a brewery. I definitely did not get rich off of that. I did learn more from that experience than any degree or course. That was invaluable from a learning experience. Be open to change and embrace it. I strongly feel that with change you will end up in a better place.

2 Dana Garves

Chemistry, BS, University of Oregon, Eugene (2010) w/ emphasis on green chemistry

Oregon Brew Labs, CEO and Senior Beer Chemist

University of Oregon, Eugene
https://www.uoregon.edu/
09/05/23

Oregon Brew Lab
https://oregonbrewlab.com/
09/05/23

2.1 College/degree pursuit

2.1.1 What inspired you to pursue a chemistry (STEM) degree?

That is an interesting question. I was inspired when I was first introduced to the structure of the periodic table and learned that it wasn't just a random organization of elements. There was a pattern there and it was fascinating to me. It made so much sense to me and the pattern on the table goes a million different ways. I had a wonderful eighth grade science teacher who inspired me and for the first time in my life was encouraging me to be a scientist. At that time nobody had said that to me. When I got to high school, I had a science elective and decided to take chemistry. I liked the periodic table and I had liked those sections of my science classes since eighth grade. It was very hard and I really struggled. I think that challenge is what interested me and I could tell that your teacher makes so much of a difference for students. I was very fortunate to have excellent chemistry teachers who pushed me. In my junior year, I was in the Science Olympiad and I was elected as president. During my senior year, they required us to have chaperones to keep an eye on us. I finally asked my mom and told them that we needed a chaperone. That was the first time that my parents found out I was into science. They were very supportive. It's interesting right, because all through high school I struggled with chemistry and flourished in all of my other classes. When it came time for me to pick a degree, I was originally wanting to do a double major in chemistry and physics. I also had an excellent physics teacher during my senior year.

https://doi.org/10.1515/9783110798777-002

Science Olympiad
I was chair of our Science Olympiad for a couple of years. This is a very exciting event for young scientists to attend. They specialize in a variety of areas including ecology, wheeled vehicles, meteorology and optics. There is a national competition every year that winners of regional competitions can advance to from.
https://www.soinc.org/
09/05/23

2.1.2 How long did it take you to graduate? Were there any stumbling blocks you'd like to share?

It took me 4 years. Honestly, I really loved chemistry and physics. Physical and inorganic chemistry were my favorites. Organic chemistry was really difficult for me though. One of my stumbling blocks with getting into the industry was I really wish that I had taken a biochemistry or a biology class or something with microbiology. All of the yeast and bacteria testing I do now was self-taught. Oregon State has a 1 week beer science course geared toward industry that I took back in 2013. I have formal training from Thomas Shellhammer, but I do wish that I had taken a bio or micro credit and diversified my classes.

OSU One Week Fermentation Program
The formal title of this certificate is "Beer Quality and Analysis Series: Microbiology and Beer Analysis" and it teaches a variety of topics, including ASBC microbiology methods, isolating yeast and contaminants and QA/QC methods.
https://workspace.oregonstate.edu/catalog-page#certificate-programs
09/05/23

2.1.3 Did you do any research as an undergraduate? If so, what was the project?

I did. I worked closely with Julie Haack doing green chemistry and disseminating the principals into their chemistry curriculum. We worked hard to bring it outside of Yale and other colleges and we worked hard to disseminate information on how to make your labs cleaner and safer for technicians. I have a rich background in green chemistry.

Green chemistry
The ACS green chemistry initiative has three primary principles: maximize resource efficiency, eliminate and minimize hazards and pollution and design systems holistically and using life cycle thinking. If you navigate to the design principles tab you can download a booklet that covers these principles.
https://www.acs.org/greenchemistry.html
09/05/23

2.1.4 Do you have any words of wisdom for students wishing to pursue a chemistry (STEM) degree?

Stick with it and it will pay off! You do not see it until you go back and look at your general chemistry books. We are given all of the knowledge of chemistry up front and it is parsed out. You do not learn about anything small like an electron until years later and you realize what an impact it has on chemical behavior. Chemistry is both small and large and there is something very beautiful about how an atom interacts with another atom. Chemistry is everywhere and you can apply it to everything.

2.1.5 How often do you use what you learned as an undergraduate in your current job?

Almost on a daily basis. I work at the lab bench and try to set aside a "meeting day" or a time when I have to be in front of a screen all day. I still do titrations and use an AA or a spectrometer. These are some of the reasons that I stayed in the industry – because I am using chemistry every day.

2.2 Movement to brewing field

2.2.1 What was your "entry beer" into enjoying craft beer? Why did you like it?

I actually have two. My father was a craft beer drinker. He liked Fat Tire and the rest of the family liked Miller. So, he was the beer "snob" of the family. Miller was always around and I tasted it and did not like it (at the time). That got me into thinking about beer, so my father attempted homebrewing, and I wanted to be a part of it. The beer that I drank that was my true "entry beer" was the Obsidian Stout from Deschutes. If the bar did not have that then Mac and Jack's African Amber was my go to. I liked the full-bodiness of Obsidian and in comparison to the roasty notes of Guinness I liked the smoothness of Obsidian.

Obsidian Stout, Deschutes
I've also really enjoyed this beer on a few occasions. It is a 6.4% ABV stout made from two-row, chocolate and crystal malts to name a few. It is described as having "distinct notes of espresso."
https://www.deschutesbrewery.com/beer/obsidian-stout/
09/05/23

2.2.2 Did you get started with homebrewing?

I suppose so. My father had a few failed attempts at brewing in the bathtub. In college, before I was at the legal drinking age I tried my hand at a cider. I did not get into homebrewing until 2015 and 2016.

2.2.3 What was your first recipe?

I think it was a brown ale and I was trying to clone a Newcastle.

2.2.4 Did you ever enter a recipe into a brewing contest?

I have actually. I did okay and won second place for a Rice Lager and I have a handful of ones for Ciders. I like hopped ciders. I grow Mt Hood hops and do a dry hopped cider of that.

2.2.5 What was your biggest disaster?

Honestly, I still struggle with sulfur in my ciders and aging them out. I can throw copper in there and I would like to get to the point where I can get it resolved before the sulfur is even produced. That is my own personal demon and we all probably struggle with it. I do struggle to hit my IBU targets (but I think a lot of the science around IBU chemistry does not take into account homebrewing.)

2.2.6 Were you part of a homebrew club? Which one? How long a member?

Yes. The Cascade Brewing Society is a homebrew club in Oregon that has been meeting for about 40 years and this year I took on the role of being President. We have meetings, social events and do a quarterly brew club member's only competition where we have various styles. I've been a member since 2014!

Cascade Brewing Society
This homebrew club has been around for more than 40 years. Unlike some homebrew clubs they specifically mention that they support makers of a wide variety of beverages including cider and kombucha. They host a camp out and club only competitions.
https://cascade-brewers.com/
09/05/23

2.2.7 Did you ever brew a beer for an event (wedding)? How did it go?

No, unfortunately. That is the thing – I am not a brewer. I do not like the cleaning. I love going for collaboration brews for an event, but I generally do not brew commercially. I have a list somewhere of all my commercial brews, even. I am still working on my skills as a homebrewer (though I prefer to homebrew hopped ciders.)

2.2.8 Prior to getting into brewing what else did you do after graduating?

I went to a water quality control company and it was a terrible job. I did not feel passionate about it, even though I was using my degree. I was using my general chemistry skills on a daily basis: running titrations, taking pH measurements and using separatory funnels. It was kind of perfect at the time because the joke was that people who were graduating weren't going into fields that used their degrees. I did feel fortunate, but I did not like the work. At the time, I had someone else in my graduating class that was interning at a local brewery. I thought that was much cooler than analyzing water. So, I looked for breweries that were hiring and there was one actually. It was a two-line Craig's list ad that said "Brewery seeking chemist to test for oxygen." I applied to that ad and did some digging and found that there was only two breweries in town that would have a QC program, and one of them was the one that my buddy worked at so the other one had to be it. I went to the Ninkasi Brewing website and found a job listing that indicated they were looking for a chemist. I applied to the website and gave my resume to the person sitting at the front desk and told them that they were going to give me this job because I wanted it. They did give me the job. I built their laboratory from the ground up including their sensory program. I stayed with them for about 4 years until 2014. I culminated my experience there with a project where we shot a beer yeast into space on a rocket. You might have heard of the beer, it is called "Ground Control." It was an imperial barrel aged stout made with "space yeast." After that I started asking myself where do I go from there. You know, I had built their lab from the ground up, actually twice through an expansion. I did not feel like I could go above that. Around that same time, I had opened up our lab to all of the breweries in Oregon. They would bring me a case or a bomber (22 oz bottle) and I would test the beer for alcohol, IBU, micro-testing, whatever they needed. Around that same time of the space launch my workload was actually 50% Ninkasi and 50% other breweries. The owners were really community-based. At first it wasn't a problem, but then when I started dipping over to 51% other breweries and 49% Ninkasi, it starts to be a problem. I thought, "why are all the breweries coming to me, aren't there other labs that could help them?" At the time it was just White Labs and Brewing and Distilling over in Kentucky. I realized there was not a lab in Oregon. With the blessing of the Ninkasi owners, I left and started my own lab. They gave me unemployment so I could get started and created Oregon Brew Lab. Now, I was doing what I did at Ninkasi except

I was getting paid to do it. I even started doing things outside of brewing to diversify my clients. So, that is sort of my origin story if you will.

Ground Control, Ninkasi
How cool is that! Ninkasi made a beer from yeast that was on a rocket. Ground Control is an Imperial Stout packing a 10.0% ABV and is made from Munich, Peated and Black Malts in addition to honey, cocoa nibs and Oregon hazelnuts.
https://nsp.ninkasibrewing.com/the-beer.html
09/05/23

2.3 Current position

2.3.1 What are your current job responsibilities and tasks?

I like to say that I am CEO and Senior Beer Chemist at Oregon BrewLab. I have been running it since 2014. I test for the basic things in beer that used the least amount of chemicals possible because I built my lab in my garage. In order to do that I prefer to have safer chemicals and that goes with my green chemistry background. I work with over 600 breweries, cideries, meaderies, kombucha and nonalcoholic beverage producers. I also like to do analysis for non-fermented beverages. I am starting to include things like coffee, tea and hot sauces. I run the whole company. I provide alcohol testing. I have an Anton Paar Alcolyzer DMA and Alcolyzer so I run all of my samples through that. It is the first test that I run in the morning. I bring the samples up to 20 °C because they have been refrigerated and then filter or centrifuge depending on their "milkshake" status. It has been interesting to have to adapt to the haze and milkshake craze. I try to keep up on industry trends. I am testing for interesting things now like polyphenols along with IBUs because it kind of gives you a little more idea about the sensory bitterness. Turbidity is something that I have recently added because of all the haze testings. I also do color, pH and titratable acidity. Additionally, I have enzymatic analysis for testing for acids and sugars. I run most of the tests in the lab that are niche such as foam collapse rate. I had a local university develop a foam collapse vessel which was expensive but was worth it. Interestingly, malt companies were the ones that were most interested in doing that particular test. Maltsters were doing hot steeps on their malts and carbonating it. They were interested in seeing how this translated over to SMASH beers (single malt and single hop).

2.3.2 What is your favorite and least favorite analysis to do and why?

I like the process of putting beer samples into the Anton Paar. It is the information that brewers care the most about. Many of these other things are interesting projects or "oh no" moments such as finding lactic acid in something and they want to know how much. I think seeing the alcohol content and how many calories are in it are something that everyone wants. The chemical process of distillation for vicinyl diketones is also pretty neat because it is basic chemistry. It is your classic general chemistry principles. My least favorite is IBUs. I have worked hard to keep my lab a "green lab" and iso-octane is my most dangerous chemical. It is my least favorite to work with. I have a micro-hood that helps protect the samples, but it does not protect me. I dress up with the full goggles, face mask and the works as well as gloves. It also does not provide the information that people think it does since it does not tell you all the bitterness in the beers. For example, if you have barrel aged your beer then the tannins from the barrel can add bitterness to the beer. Those are not accounted for in IBU assays. It is my most dangerous test as a green chemist and it is probably the most misleading. What I also don't like is that opening up so many beer cans in a day can wreck my fingertips. So, I use a barback can opener now.

2.3.3 What classes that you took apply to some of the things that you have done in the brewing industry? Why/how?

I followed the basic chemistry track and while I was there I dove heavily into the green chemistry part. I really loved inorganic and physical chemistry. Those were my favorite. I wish that I had cared more about organic chemistry and that I had taken a biochemistry course. That would have rounded out my knowledge for the beer industry. I did not have a knowledge of things like the ATP cycle or even just fermentation or microprocesses for that matter. I had to learn all of that within the industry instead. I was fortunate when I worked at Ninkasi that I could go down to OSU for a week-long analysis class and get the college version of how to handle yeast, and how to handle microsamples for *Lactobacillus* or *Pediococcus*. I wish that I had gone more into the bio with my elective courses during college.

2.3.4 What adjustments did you need to make from working in a lab setting to a brewery setting?

I would say that the biggest transition for me was that when I talked to another scientist about error, margins or variations – this is us expressing the potential range of results. When you talk to a brewing person who does not have that background they are turned off by things like "error" and all they hear is "the result is bad." It took a year or two for me to understand how others view scientific concepts like error. Tracking your margins of error are good practices for a scientist of course, but when you are talking to somebody about whether to dump a beer all they here is "error equals wrong." We also

like our long significant figures as scientists, and beer only carries out to a single deci-
mal place or maybe they only really care about the whole number.

2.3.5 What do you enjoy most about your current position?

Like most entrepreneurs, I like not having to answer to a boss, but the other side of that
is that my customers are my bosses. In my position though I can pick and choose as to
who those people are. I love this industry. In terms of being a woman it has been a little
extra hard, but I have felt fortunate in that it has been met with mostly positive reactions. It is sometimes shocking when I go to other areas of the country and discuss my
experiences.

2.3.6 Do you have any words of wisdom for students wishing to apply their chemistry (STEM) degree toward getting into the brewing industry?

Hang out at breweries, go to homebrew club meetings and immerse yourself in it. Many
scientists are able to get into the industry with our background. There is an artistry behind brewing and that is the hardest thing for a scientist to learn. It is easy to make a
good technical beer, but that is not what sells or excites consumers. The best way to be a
well-rounded scientist entering the industry is to have a love and passion for the industry. Do you have a favorite beer or brewery? Talk to the brewers and bartenders when
they are grabbing their off-shift pint.

2.4 Experiences in the brewing industry

2.4.1 How would you describe your leadership style?

I am pretty collaborative. I will sit back and let everyone speak their peace and get an
idea of what the overall consensus is. I will then paraphrase that back and see if that is
sitting right with everyone and the way that I heard things and then move forward
with a more diplomatic approach to a problem or a solution. One of my favorite parts
of the job is troubleshooting. When someone comes to me with a problem and asks me
if I can solve it, I like that. It is not just putting it into an instrument. I would say that
my style is quite a bit of troubleshooting and what I call MacGyvering. It involves taking
what I have around me and then using it to solve a problem.

2.4.2 What are some unique strengths that you think women bring to the brewing industry?

I think that what they bring is a different perspective. Women have a better palette and can be more descriptive in that area. They can be a key part of your taste panels. Any time that we have more types of people in an industry the industry will grow. It extends to all types being involved and helping things become better. When a person sees somebody who looks like them drinking a beer then they want to. Also, what we need in the industry right now is more beer drinkers. It is a little scary to see the number of beer drinkers dropping. Anything we can do to bring more people in even if that means adding watermelon puree to your IPA. So long as you still brew that Belgian Trippel for me – do what you have to do to bring more people in.

2.4.3 Have you had any negative experiences in the brewing industry that you would like to share?

Yeah I actually have, but the nice thing is it is over-shadowed by all of the good ones. I had a journalist who kissed the top of my hand when I extended it to shake hands. That was one of my most negative experiences. Also, I try to stick to similar guidelines that BA uses for labels or names of beers. If I see something that makes me uncomfortable, then I send them an email and tell them that this beer label would not be accepted at the Great American Beer Festival (GABF) and it won't be accepted at this lab. If it makes me uncomfortable, then it will make others as well. I try to take those moments as isolated events because everyone gets a chance to learn.

2.4.4 Do you have any advice that you would give to your younger self before getting into the brewing industry?

Take a microbiology or biology class. I worked at one brewery and one QC lab and I wish that I had taken a job with another regional/national brewery and I did not take it because I did not want to move across the country. I kind of wish I had taken it so I knew what a lab at a brewery of that size would look like. I wish I had taken that opportunity, but at the same time I was able to shoot yeast in space and start this company. I'd say get experience at a few breweries first. It is okay to move around in this industry.

2.4.5 Do you have any tips or tricks for maintaining a positive work-life balance?

Yes. Get a work phone and shut it off when you are not at work. That is all that I got. Do not check your email when you are off. It is so important for your body and your head that you take breaks from work. Especially in this industry, when you do get off, you sit down with a beer and the next thing you know you're evaluating it on a sensory basis. We do not technically leave the industry as a result. When I got a second phone for my work like it changed my weekends.

2.5 Professional affiliations

2.5.1 What current professional organizations are you affiliated with? Why did you join them?

I have been a member of ACS since I went to a conference during my college years. Sometimes I let it lapse though. I have participated in ACS workshops and did some things with the Portland chapter. ASBC is another one and I joined because I use their methods and like the forum and going through that to see what labs are having trouble with. I am in the MBAA Northwest Chapter and I serve on the technical committee. I am also on the DEI committee for the Oregon Brewer's Guild and am in three chapters for the Pink Boots Society. I am president of the Homebrew Club I mentioned earlier (Cascade Brewing Society) that has been in Eugene since 1982 – we just celebrated our 40-year anniversary. I am a member of the BA when I judge GABF.

Oregon Brewer's Guild
This Guild has existed since 1992. Their mission is to "promote and protect the state's brewing industry and the common interests of its members." They host Zwickelmania (sampling port) every year where members of the public are invited into breweries to learn more about brewing, food/beer pairings and the industry as a whole.
https://www.oregoncraftbeer.org/
09/05/23

2.5.2 What do you enjoy most about going to their events?

The networking and getting to see beer scientists from other parts of the country whom I have not met in person yet and have a pint with them. It's great to hang out with somebody who understands what you do. It's great to be face-to-face with industry peers. I keep a detailed notebook for my workshops or sessions and come back to pull out key points that I want to explore more deeply so that I can look at notes from previous years. You are constantly relearning things throughout your career and that should be applied infinitely if you work in a scientific world.

2.5.3 How important is it to be associated with other professional organizations?

When I was going to school I did not really understand what a networking group did or how it worked. That was probably an age thing. I would say that they are very useful in terms of generating your own network or sphere of people who know you to be a good scientist or resource. You can show people that they can turn to you. Many people go into networking thinking that it will help them so much. When you offer to be a helper that network explodes for you. Pick one to start with and get it on your resume. If you have been a continued member it means more and it helps you become a better scientist or brewer if you're already in the industry. MBAA is one of the top ones and ASBC is fairly niche. The technical forums on MBAA are also unmatched in this industry.

2.6 Other/miscellaneous

2.6.1 Outside of brewing what other hobbies or interests do you have?

I am a gardener. We have a small plot of land that is full of fruit and veggie beds. We grow tomatoes and hot peppers. We don't just do jalapenos – we have Trinidad scorpions, Carolina reapers and Habaneros. We also have a blue dragon's tongue and it may ruin our whole bed. I am worried that it will cross-pollinate with the other peppers – it is a 2 million Scoville pepper. I also paint with acrylic on canvas and get some creative outlets that way. When the weather is nice we like to go camping and hiking.

Scoville scale
This scale started off on an organoleptic basis where five "lucky" individuals tasted dilute solutions of a pepper solution until they could not register the heat of it. The score was based on the dilution level required. Thankfully, this can now be done on instruments, namely HPLC using capsaicin as the standard. Pepper ratings range from above 2 million all the way down to 0 for bell peppers.
https://www.alimentarium.org/en/story/scoville-scale
09/05/23

2.6.2 Is there anything else that you'd like to share about your personal or professional life that you think people might be interested in knowing?

I think that if you are able as a student to identify your strengths and what you are good at it, you can carry those things with you through your whole career. I like to hang on to the environmental aspect and I like beer and how can I do those things together. If you identify that earlier, then so much the better.

3 Sean Greenwood

BSc (Hons) (MS equivalent), Chemistry, University of Liverpool, UK (1990)

Transport Brewery, Shawnee, KS, Brewer

University of Liverpool
https://www.liverpool.ac.uk/
08/23/23

Transport Brewery
https://www.transport
brewery.com/
08/23/23

3.1 College/degree pursuit

3.1.1 What inspired you to pursue a chemistry (STEM) degree?

My father was actually a chemist. When I was at school I think some of that rubbed off on me when I was in middle school and was trying to decide what I was going to do. The education system is different in England. You do general education subjects until you're 16, and then you specialize for 3 years on three subjects which would likely be what you would study in college. They would be equivalent to AP classes over here. So, I did what you would consider the sciences: math, physics and chemistry. Those were the more specialist subjects. I then decided that chemistry was what I was more interested in as opposed to the more mathematical side of things such as physics.

3.1.2 How long did it take you to graduate? Were there any stumbling blocks you'd like to share?

In England, you can finish in 3 years. You do not do the additional general education for 1 year that they do over here. For stumbling blocks, you just need to remember that learning things is hard. I went to a boarding school for high school and I was used to being independent and living away from home. My oldest daughter just started as a freshman, and it was a bit of a shock for her to be independent and have to look out for herself and setup her own schedule. I was actually used to doing that since I had been away from home during my middle and high school life. That was less of an issue for me.

https://doi.org/10.1515/9783110798777-003

3.1.3 Did you do any research as an undergraduate? If so, what was the project?

It was a long time ago. We were looking at α-pinene epoxides and opening them up with nucleophiles. We studied the mechanism of how that took place and my professor was interested in the variety of ways that it could happen. It was pretty basic stuff. He had a team of post-docs and graduate students who did the main part of the work. So, a lot of these projects were just to teach you the general techniques and stuff like that. I am not sure if we managed to discover or prove anything.

α-Pinene epoxide
These are very reactive and are prone to side reactions that decrease the yield of desired products. One derivative, camphonelic aldehyde, is used in fragrances.
https://www.sciencedirect.com/topics/chemistry/alpha-pinene-oxide
08/23/23

3.1.4 Do you have any words of wisdom for students wishing to pursue a chemistry (STEM) degree?

It is very different over here as well. It was a long time ago. I would say to stick with it. Chemistry is one of the more difficult subjects that is intellectually demanding. Stay with it because the rewards for working in it can be very rewarding. You are helping people and actually doing some good in the world.

3.1.5 How often do you use what you learned as an undergraduate in your current job?

All the time. I still have my old textbooks. I do not tend to look at my written notes anymore, but I still have all those books, you know, the general textbooks. They are probably obsolete now, but the classical part of chemistry never really changes. In fact, I was looking at an inorganic chemistry textbook today and I bought that book over 30 years ago.

3.2 Movement to brewing field

3.2.1 What was your "entry beer" into enjoying craft beer? Why did you like it?

I came into it from a different angle of course. There has always been small breweries in the UK. When I first started drinking beer, in the UK, the most popular beer for young people was imported German lager. I did not really get interested in those. I was more into the traditional bitters and milds, and my friends thought those were old man beers. I liked traditional cask conditioned beer. It was always available, but it just was not seen as a modern beer. I then joined the campaign for real ale and went to festivals and went around our local area. We would do surveys that were used to generate a beer guide. It was a tour guide for pubs that would catalog the beers that they sold, whether they had beer engines and cask conditioned ales or whatever. I was involved in all of that until we moved to America.

We ended up moving to Kalamazoo with Pfizer. Initially, we moved over here originally for just 2 years and ended up staying permanently. At that point, Kalamazoo in 2006 Bell's Brewery was up and running. They didn't have a small craft beer scene at the time though. We would go to Bell's Brewery a lot and we were intrigued by Bell's. I think it was Two Hearted Ale. It was the first time that I tasted a heavily hopped beer or America style ales. Sierra Nevada Pale Ale was another one that was unique to me at the time. Two Hearted was an interesting beer. English Ales were similar to each other at the time. Nobody was going out or experimenting with unusual flavors or the like. We were not used to those hop varieties in England such as the West Coast hops. The backbone of the beer was an English Ale with crystal malts and traditional grain bills. It had some unusual flavors as well.

Two Hearted Ale
This 7.0% ABV beer is brewed at Bell's Brewery and uses Centennial hops as a dry hop. The beer was named after fishing waters in Michigan.
https://bellsbeer.com/project/two-hearted-ale/
08/23/23

3.2.2 Did you get started with homebrewing?

Yes. That is how I got started.

3.2.3 What was your first recipe?

We used to have a party every year and buy a couple of kegs of beer. I knew a whole bunch of homebrewers and so we decided to have a homebrew beer festival for the party instead of having to buy it. Somebody asked me if I was going to make any beer and I said no since I had not made any beer yet. They said that I should, and that it is not any different than what we were doing at Pfizer. So I jumped into it and ordered an extract kit from Northern Brewer. It was a turkey fryer-type kettle and a couple of Rubbermaid mash tun and hot liquor containers. I bought a book from the campaign for real ale called "Brew Your Own British Ales at Home." I was missing those and so I think the first one I made was a London Fuller's Pride clone. The first one or two batches I did were extract and then after that I was all in and went to all-grain. I was hooked and wanted to go as complicated as I could.

3.2.4 Did you ever enter a recipe into a brewing contest?

I've done a few actually. I have medaled in quite a few competitions. I still do home-brew a little bit and submit beers once every few months. This year, I entered a couple of competitions with a best bitter and an ordinary bitter and a Helles Exportbier. I ended up medaling with a silver and a bronze for the bitters. I got a gold for a Czech pilsner and a gold for an Australian Spartan ale. I'm in a couple of clubs locally and so we can enter through the club and that makes it easy.

German Helles Exportbier
The BJCP ABV range for this beer is between 5% and 6%. It is described as a "golden German lager balancing a smooth malty profile with a bitter, hoppy character in a slightly above-average body and strength beer."
https://www.bjcp.org/style/2021/5/5C/german-helles-exportbier/
08/23/23

3.2.5 What was your biggest disaster?

I was brewing one day in the middle of summer and our A/C system broke and I thought that I was not going to pitch yeast until it got cooler. It took 3–4 days for the engineer to come out and get the A/C back on. I should have just bought some kveik and be done with it. The thing started fermenting and I let it ferment out and it had a terrible infection and was a drain pour straight away.

3.2.6 Were you part of a homebrew club? Which one? How long a member?

When I was in Kalamazoo, I was not part of a club. Once we moved to Kansas though I was out of work for a bit since we moved for my wife's job. I thought it would be a good way to meet some new people. I have probably been involved for 6–7 years. Initially I was in the Johnson County brewing society in Kansas City. I am also in the Kansas City Bier Meisters club. They are a big club in the area and won Homebrew Club of the Year from AHA and some of their members have won gold in the national competitions. They are definitely a big club. There is a difference between the two. The Bier Meisters are very technical and give lectures on technique and the technical side of things. Johnson County is where guys would get together and taste and talk about the beers they brewed. A little more social.

Kansas City Beer Meister's Club
They are the oldest homebrew club in Kansas or Missouri, founded in 1983, and they did indeed win Homebrew Club of the year in 2022. Aside from monthly meetings, they host several competitions and events throughout the year. The club has over 100 members and emphasizes providing members with good feedback on their beers.
https://kcbiermeisters.com/
08/23/23

3.2.7 What was your biggest batch as a homebrewer?

I just did 5 gallons on a homebrew scale with the setup I have. I have a single keg system. I initially had a Robobrew (now called Brewzilla) and now a Clawhammer system. You know a brew in a bag-type system. After doing the turkey fryer out in the snow, I wanted to be able to move inside.

3.2.8 Did you ever brew a beer for an event (wedding)? How did it go?

No, but I have done beer festivals where we would give away our beer as part of them. I ended up taking four beer engines to one and serving cask conditioned beers.

3.2.9 Prior to getting into brewing what else did you do after graduating?

I worked at Pfizer in the UK doing organic chemistry. We were doing drug discovery. We were synthesizing different analogs and screening them. Then we were working on the process side of that. I have a couple of patents for process chemistry. We had a drug called Simparica for dogs that is used in dogs for killing ticks and fleas. They had a facility that worked on animal health. At that time (1990), they were not only looking at human drugs, they were also looking at animal health. Of course, with animal health drugs, they get to market much quicker. You see more of an immediate impact with the animal drugs. Originally, they were looking at farm animals and then they started to work with pets as well. They primarily worked on antiparasitic compounds such as ticks and fleas and worms. They started with sheep, pigs and cows and then realized that there was money in treating cats and dogs so they transitioned to that.

3.2.10 Would you recommend brewing to other chemists out there and why?

So I am involved in a very small brewery. It is a lot less professional than someone who wants to get into the business and make it as a career. For example, we do not

---

Simparica
Simparica is produced by Zoetis which evolved from the animal health division at Pfizer and is now a standalone corporation. This monthly drug is used to "protect dogs from ticks and fleas for 35 days." Why 35 days? So that owners have a few days of grace to dose their beloved pets of course.
https://www.zoetispetcare.com/products/simparica
08/23/23

currently take a salary in the brewery. The three owners do it as more of a labor of love. We hope to get money on the back end. I can't really recommend it as a career given our particular situation, because it is not a career for me.

3.3 Current position

3.3.1 What are your current job responsibilities and tasks?

In my current, real job I work as part of a synthetic chemistry team for a local research institute. We have a variety of clients where we contract for. We do discovery, drug development, Good Manufacturing Practices and synthesis. We have industrial and government contracts as well. We do some pharmaceutical work too. So we do a variety of things which is different than just working in pharmaceuticals where I did not get as much exposure to these varied activities. That's good and I've only been doing that for a year. Before that I was actually using my chemistry knowledge as a regulatory affairs consultant. We were looking at post-approval drug dossiers. If there was any change to the manufacturing procedure, then the chemistry manufacturing and controls dossier has to be kept up to date. So if there is any change to the process or the site for the manufacture of drugs where a site is shut down or if the analytical methods that they are using change over time, all of those things need to be written up and submitted to all the relevant drug authorities around the world. Here, we have the FDA and every country around the world requires these dossiers so they have to be submitted to them. I was doing that and that was sitting in an office working on spreadsheets and word documents. Brewing was my escape from that. Then when Covid hit, I realized I did not want to do that anymore. I was working from home and felt isolated. I quit that job and this new chemistry job came up. I was working at the brewery during the week in between jobs which was good. Prior to that with the four of us brewing we were working weekends as well. It took some of that pressure off everyone else then. As far as my responsibilities at the brewery are concerned, I am one of five owners of which three are hands on in the brewery. While we don't have a head brewer, I am probably the one that is closest to that. I would do recipe design for recipes. We have an assistant brewer during the week who handles things now. I am doing hands-on brewing, sampling and cleaning tanks. You know, all that stuff. Also doing kegging. I do not do ordering or raw materials since others do that. I am really more involved in production and kegging. We don't have any distribution beyond what we have in the taproom so we don't have to worry about canning which is nice. We have two three-barrel brews in a bag system. The company was called Colorado Brew Systems and they have gone broke now. Their system is called Dual Nano brew I think that is what it was called. It has two separate three-barrel kettles so we could do six barrels at a time if we doubled up. We have two three-barrel fermenters and three six-barrel fermenters and a three-barrel bright tank. We are a small operation now although we are about to open a second location in a town nearby about a 20 min drive south. That will have a three vessel more professional system that will be ten barrels. To start with, we will have five or six ten-barrel fermenters and we will be able to distribute between the two locations so we will be able to do our stock beers on a bigger scale and push kegs from one place to another. We will still have the small system to do experimental or one-off brews. We may do all of our cask beers that way too.

Cask beers and beer engines
First time I heard the term "beer engine" I could not help but think of a car that is powered by beer. While it sounds very complicated, a beer engine is essentially a hand-powered pump used to dispense beer from a cask. There is no force carbonation used in beer engines so what carbonation that does exist is naturally produced.
https://www.beveragecraft.com/blog/beer-engine-guide
08/23/23

3.3.2 What classes that you took apply to some of the things that you have done in the brewing industry? Why/how?

Not a whole lot really. A lot more of it is biochemistry and microbiology with yeast, and I did not take any biology classes. I was doing the classical organic and inorganic and physical chemistry. The experience of working in a lab though was helpful being used to working in large liquid volume environments, moving liquids around and working in stainless steel and things like that. That translated more to what I'm doing on the brewing side now. When I was at Pfizer we had a fermentation group so a lot of the animal health drugs back then were made by fermentation.

3.3.3 What adjustments did you need to make from working in a pharmaceutical setting to a brewery setting?

I think the biggest difference is the sanitation side of things. You know, things don't grow in a chemistry setting. So, you just have to worry about oxygen, but you don't have to worry about keeping things that clean or sanitation. Things are on a bigger scale in the brewery. In a lab you just have glassware that you throw in the sink and give a good scrub to. Moving huge volumes of liquid using pumps is different. It is a similar step up from homebrewing to the scale of industrial brewing.

3.3.4 What do you enjoy most about your current position?

I enjoy experimenting with things and making things. That is what I always enjoyed about chemistry as well as creating things. The good thing about the brewery is I can walk into the taproom and visit with customers about what they like. I can get some good feedback from them about what is good and even what they did not like because they will tell you that as well. It is nice to create something and get feedback on what that you made.

3.4 Co-owner and brewer

3.4.1 What drew you away from pharmaceuticals to the brewing industry?

I left it because we had to move house. I always assumed coming to Kansas City, where there is a big animal health sector here, that it was going to be easy to get a job. As it turns out, many of the companies here are actually start-ups so they would just contract out their lab chemistry. I ended up having to do an office-based job. I got more into homebrewing then and started to enjoy it more because I was missing the hands-on aspect of working in a lab and experimenting with the things so it satisfied that. I actually got into the professional brewing area through knowing some people from the homebrew club. The people who owned the brewery already knew me. For their first anniversary, they asked me to come in, design a beer for them and help them brew it. They wanted me to lend them my beer engine since they did not have one and so we did an English cask conditioned ale. That was my first taste of the larger scale side of things and I really liked it. I thought it was pretty cool. About a year later, the head brewer decided to leave and they offered me the chance to buy in to the ownership team. It was because I knew the people through the homebrew club and they knew me by that point. They would still show up at the homebrew meetings because, of course, that was how they got started.

3.4.2 What is the biggest challenge to brewery owners both locally and nationally?

Currently we are finding it harder to sell the volumes that we were selling previously. I think the economy has changed some, and so it stops people from going out, inflation. Then our cost of goods has gone up too due to worldwide issues. That is difficult now and then you have people who do a "dry January" so we have that coming up as well. We do have a great marketing person who helps to drive sales. We come up with novel things to bring people in. It is a tough time to open a brewery. Working on our new facility before Covid we had quotes to do everything and now we find out that the costs have doubled due to labor or supplies. We have had to put in more money or take out bank loans to cover the whole thing. At the moment, it is probably not a great environment for opening up any small business.

3.4.3 Do you have any advice for people who think they want to own or start a brewery?

Do not think that it will make you any money initially. If you are going to do it then try to have another means of income whether it is a second job or money is saved up. It is a tough industry to make a living. Many are thinking about doing it as a retirement situation where they have a pension from their previous job. Many people will struggle if they want to do it as their sole source of income.

3.4.4 Describe your toughest day in the brewing industry so far

My toughest day so far was a cleaning day. We bought a lot of used equipment and one of those things was a heat exchanger with 100 plates in it. It is very large like a couple of feet tall with huge bolts that hold everything together. We were having trouble with batch quality and we wondered if it was due to contamination in the heat exchanger. My wife dropped me off at the brewery and I originally told her it would be finished in just a couple of hours. This thing took me 8 h to clean. I took the pieces out and it was caked in with trub and hop debris. It was concreted on. You have to then put it back together and make sure that everything fits right. You have a gasket that goes between the plates and I pressure tested it and it was leaking. We had a brew day the next day so I had to get it fixed. I see on our calendar that it is due to be done again and luckily I don't have to do it this time. Stuff gets trapped in there so I watched a video so I could show the other guy how to do it. I was alone when I did it last time so nobody saw the horror of dealing with it. The guy who did it in the video had the same type of stuff in there as we did though – it is just amazing.

Heat exchangers

There are many options available to cool wort off. Some of the more interesting ones for homebrewers are sticking the brew kettle in a snowbank or carving out a block of ice to set it on. Heat exchangers are another option and utilize water and metal to draw away heat from the wort. These are not much different than the radiator you find in a car except that they are smaller and do require periodic cleaning due to the collection of brewing debris (e.g., trub) in the metal parts being used to transfer heat to water. There are four ports on these: two for wort in/wort out and two for water in/water out.

3.5 Professional affiliations

3.5.1 What current professional organizations are you affiliated with? Why did you join them?

I was in the Regulatory Affairs Professionals Society called RAPS just because I was doing regulatory affairs. I let that lapse since I am no longer doing that for my current job. Most recently, I joined ASBC like a month ago. We went to a local meeting and there were some people there from Kansas City Brewing Company and some of the local breweries. They are large enough to have some to have chemists associated with it. I looked into joining before because being the only chemist in the brewery I am going to be the one who understands and is responsible for that type of stuff more than the rest of the guys. We decided it would be a great investment to have all of that available. Certainly when we open the new brewery we will have room to have a small lab to do some yeast management and to analyze for infections. It will be nice to have that access to methods now.

Regulatory Affairs Professionals Society

This organization, founded in 1976, provides support to individuals "involved with regulatory and quality for healthcare products, including medical devices, pharmaceuticals and biologics, diagnostics, and digital health." Resources for members include educational materials, certification opportunities and career networking.
https://www.raps.org/who-we-are
08/23/23

3.5.2 What do you enjoy most about going to their events?

I literally only joined a month ago so I cannot say much about ASBC events.

3.5.3 How important is it to be associated with other professional organizations?

I think it depends on what other access you have. When I worked at Pfizer they had databases that they paid for and that we could use and instantly get journal access. If you are working for smaller universities or companies, then you may not have access to that vital information. Joining those are very important in that case.

3.6 Other/miscellaneous

3.6.1 Outside of brewing what other hobbies or interests do you have?

I like to listen to rock music. I do not have a lot of time since I work full time and the brewery is open on the weekends. I used to do snowboarding when I was younger. Coming from England, I like to watch rugby and soccer. I also like spending time with my family. The brewery side is all encompassing. If I am not brewing then I am thinking about homebrewing or reading books on the history of beers and things like that. I am a bit of a brewing geek actually.

3.6.2 Is there anything else that you'd like to share about your personal or professional life that you think people might be interested in knowing?

As far as the professional side of things we are the only ones doing English style or cask conditioned ales in the Kansas City area. The week before Thanksgiving we had a beer festival where we had six different cask conditioned beers all at once. It was unbelievable the numbers that showed up for it. We had a line out the door for 3 h. Before that we would do one a month and people did not know what to do with this warmish beer we were serving. Now, people will come from neighboring cities specifically because we have that now, so I am excited to be able to do the education part of that. With the new facility we are excited at the prospect of being able to have a permanent cask beer on hand all the time as one of the standards that we offer. I am excited about that and being able to educate people on traditional English styles of serving and conditioning and things like that.

4 Terry Kellar

BS, Chemistry, University of Pittsburgh, Pittsburgh, PA, 1994

MS, Organic Chemistry University of Pittsburgh, Pittsburgh, PA, 2006

El Cid Brewing Company, San Diego, CA, Owner/Brewer

University of Pittsburgh
https://www.pitt.edu/
08/23/23

El Cid Brewing
https://elcidbrewing.com/
08/23/23

4.1 College/degree pursuit

4.1.1 What inspired you to pursue a chemistry (STEM) degree?

A very strong interest in chemistry during my first general chemistry class. I originally didn't know what degree I wanted to pursue, but I found myself reading the textbook on Friday and Saturday nights. Then I took organic chemistry, and I knew I wanted to be a chemist.

4.1.2 How long did it take you to graduate? Were there any stumbling blocks you'd like to share?

It took me 4 years to graduate with my BS, and everything went smoothly for the most part. My girlfriend became pregnant during my first year of college, so that made things a bit more challenging with time and finances.

4.1.3 Did you do any research as an undergraduate? If so, what was the project?

I didn't do any research, but I did work as a teaching assistant. My original plan was to teach chemistry at a university, but my career goals obviously changed later on. I taught the honors organic chemistry lab. I would prepare lectures each week, and we

https://doi.org/10.1515/9783110798777-004

would have discussions before and during the lab. Some of those students were very bright and knew more about chemistry than I did at the time.

4.1.4 Do you have any words of wisdom for students wishing to pursue a chemistry (STEM) degree?

Yes, I would highly recommend an advanced degree if you plan to go into the industry. A BS degree will not get you many interviews anymore, so a few extra years of schooling could really be worth it, particularly in medicinal chemistry.

4.1.5 How often do you use what you learned as an undergraduate in your current job?

As a medicinal chemist, I use it all of the time. It's also useful when troubleshooting a brewery issue. I find myself explaining brewing chemistry to curious customers as well, so I would say that I use it every day in one way or another.

Medicinal chemistry
This is an interesting career path for chemists. They have an opportunity to impact people's lives and well-being. You may find yourself extracting bioactive compounds from plants or designing drugs to address a particular medical need. The career link from ACS is certainly worth a read if this is an area you are interested in getting into.
https://www.acs.org/careers/chemical-sciences/fields/medicinal-chemistry-pharmaceuticals.html
08/23/23

4.2 Movement to brewing field

4.2.1 What was your "entry beer" into enjoying craft beer? Why did you like it?

Belching Beaver Peanut Butter Stout was my favorite beer for a long time. I still love that beer so much. It has a rich, full-bodied feel to it with just the right amount of sweetness.

4.2.2 Did you get started with homebrewing?

Yes. I was drinking specialty beers quite a bit that I couldn't find in most places. So I decided to start homebrewing the beers that I wanted to drink.

Belching Beaver Peanut Butter Stout
This is a great beer that I've had the opportunity to try on a couple of occasions. It is a 5.3% ABV stout "with cascading aromas of buttery peanuts and dark chocolate."
https://store.belchingbeaver.com/collections/beer/products/peanut-butter-milk-stout-6-pack
08/23/23

4.2.3 What was your first recipe?

It was a Session IPA, a clone of Founders All Day IPA. I wanted to make one that was challenging to me.

Founder's All Day IPA
The All Day is a 4.7% IPA with 42 IBUs is "naturally brewed with a complex array of malts, grains and hops. Balanced for optimal aromatics and a clean finish."
https://foundersbrewing.com/our-beer/all-day-ipa/
08/23/23

4.2.4 Did you ever enter a recipe into a brewing contest?

Nope. I went from homebrewing to professional brewing in less than a year.

4.2.5 What was your biggest disaster?

I was mashing in for a stout and did not realize that the strike water temperature was way too high. The temperature controller for the burner had malfunctioned. I ended up mashing in at over 180 °F, which ended up denaturing all of the enzymes. That was my first 10 gallon brew, and it was heartbreaking.

4.2.6 Were you part of a homebrew club? Which one? How long a member?

Not a club, but I've been a member of the American Homebrewers Association for over 3 years now. I try to support homebrewing as much as I can.

4.2.7 What was your biggest batch as a homebrewer?

A half barrel. Filled the boil kettle to the brim, so it was a bit scary.

4.2.8 Did you ever brew a beer for an event (wedding)? How did it go?

Yes, I did. I brewed 5 gallons of guava wheat beer for a happy hour at work. People loved it, and it was gone in less than an hour. That was a confidence builder!

4.2.9 Prior to getting into brewing what else did you do after graduating?

I was a formulation chemist for several companies before moving into the biotech industry. I've been a medicinal chemist since 2006, working in the San Francisco Bay area as well as San Diego.

Formulation chemistry
Another career path for chemists. This field involves making products using a variety of compounds that "contribute to the final product in some way." What makes it unique is that the chemist is actually striving for *no reaction* among the chemicals being used to generate the formula.
https://www.acs.org/careers/chemical-sciences/fields/formulation-chemistry.html
08/23/23

4.3 Current position

4.3.1 What are your current job responsibilities and tasks?

We have no employees, so pretty much everything. We are a small brewery and probably make about 250 barrels per year. We brew, bartend, clean, apply for permits, file taxes and everything in between.

4.3.2 What is your favorite and least favorite process to do and why?

I love mashing in, with the steam in your face and the smell of freshly moistened grain. My least favorite part is cleaning the fermenters.

4.3.3 What classes that you took apply to some of the things that you have done in the brewing industry? Why/how?

Organic chemistry helps quite a bit for understanding basic brewing chemistry as well as off-flavors. The biochemistry, of course, is very helpful for understanding yeast metabolism. Even the basic general chemistry stuff helps with unit conversions and calculating concentrations.

4.3.4 What adjustments did you need to make from working in a lab setting to a brewery setting?

If you're not a process or manufacturing chemist, the large scale of brewing feels much different. Also, when you work for corporate America, you have days off. When you own a brewery, you are always working, 7 days a week.

4.3.5 What do you enjoy most about your current position?

Bartending and talking to customers, for sure. A lot of good people that are fun to speak with come into the brewery. The homebrewers are usually interested in checking out the professional equipment in the brewhouse. It's always fun seeing them get excited.

4.3.6 Do you have any words of wisdom for students wishing to apply their chemistry (STEM) degree toward getting into the brewing industry?

I would get into the industry sooner than later. You don't need a lot of experience as a homebrewer, but you should brew a few batches to make sure that you like it. I'd also suggest attending a short, practical brewing certificate program to learn more about the industry and to start networking.

4.4 Experiences in the brewing industry

4.4.1 How would you describe your leadership style?

I would say that I am very interactive and use feedback from others to help make decisions. My business partner and I think differently about many things, so we typically have to meet in the middle.

4.4.2 What are the various ways that you keep yourself grounded and you use to take care of yourself?

I like to walk my dogs. I have two pit-bull mixes, and they have a lot of energy. It's always fun spending time with them, and they make sure that I get enough exercise. I also spend time reading. I have about 30 books dedicated to brewing that I use to put myself to sleep at night.

ASBC Books
If you're looking for brewing books this is a great place to navigate to.
https://my.asbcnet.org/asbcstore
08/23/23

4.4.3 Have you had any negative experiences in the brewing industry that you would like to share?

For me, the most negative aspect of the industry is dealing with the business side of things. Everything you want to do usually requires a permit or costs way more than you would expect. It can be difficult to get answers to legal or procedural questions, and it takes twice as long as you would expect.

4.4.4 Do you have any advice that you would give to your younger self before getting into the brewing industry?

Yes. Get into the industry while you're still young. I know a few brewers that started in their 20s, and they have so much practical knowledge. I do wish I had gotten started earlier, but it's never too late.

4.4.5 Describe your toughest day in the brewing industry so far

Unloading our 5-barrel brewing system. A truck arrived early in the morning with the equipment, and we only had 2 h to unload it. We had hired a forklift driver who, after arriving an hour late, told us that he wasn't skilled enough to unload it. We scrambled to find a driver, and fortunately a brewer down the street had some experience unloading equipment. That was an expensive day after paying extra for the fork lift driver and the additional truck time.

4.4.6 Anything else you'd like to share about brewing in the San Diego area?

One of the things that I love about brewing here is the camaraderie within the brewing community. There are around 170 breweries in the county, and the brewers are a very supportive group. We try to help each other out whenever we can.

4.5 Professional affiliations

4.5.1 What current professional organizations are you affiliated with? Why did you join them?

I have been a member of ACS for a long time, one of the best resources for professional chemists. I am a member of the MBAA, which is a nice way to stay up to date on new developments in the brewing industry. I am also a member of the San Diego Brewer's Guild, which is a great networking resource for local brewers.

San Diego Brewer's Guild
They boast that there are over 150 breweries in San Diego now. The site lists different events and has a directory of breweries. A job board and unique statistics are also available.
https://www.sdbeer.com/
08/23/23

4.5.2 What do you enjoy most about going to their events?

It has been a while since I have been to an ACS meeting, but I always enjoyed learning while drinking with friends that you haven't seen in a while. We have a monthly meeting with the brewer's guild, which is a great opportunity to meet others in the industry.

4.5.3 How important is it to be associated with other professional organizations?

I think it's very important for networking, as well as staying current on industry developments.

4.6 Other/miscellaneous

4.6.1 Outside of brewing what other hobbies or interests do you have?

I have been a motorcyclist for over 25 years. I love getting on my bike, riding to the middle of nowhere, and being by myself for a little while. It also reminds me that my adrenal glands still work.

Zen and the Art of Motorcycle Maintenance
True story. A chemistry colleague of mine had one of their dissertation committee members actually assign them this book for their qualifying exams. Sound weird for a PhD candidate to have to read something like this? Then give the book a read – you'll hopefully understand why it was assigned.
https://www.amazon.com/Zen-Art-Motorcycle-Maintenance-Inquiry/dp/0060589469
08/23/23

4.6.2 Is there anything else that you'd like to share about your personal or professional life that you think people might be interested in knowing?

Nope. I'm not that interesting anymore. I feel like I would've had a much better answer for you 10–20 years ago!

5 Ricky Kremer

Coe College, Cedar Rapids, IA 2010, BA in Chemistry

Singlespeed Brewery, Waterloo and Cedar Falls, IA, Lab Technician and Cellar Lead

Coe College
https://www.coe.edu/
08/25/23

Singlespeed Brewery
https://www.singlespeedbrewing.com/
08/25/23

5.1 College/degree pursuit

5.1.1 What inspired you to pursue a chemistry (STEM) degree?

Went to Coe College and majored in chemistry. I came in as a freshman and that was what I wanted to do and walked out as a senior with that very degree.

5.1.2 How long did it take you to graduate? Were there any stumbling blocks you'd like to share?

I spent 4½ years at Coe. That extra half year was spent on student teaching. While I was there I worked at a summer camp. It was a camp that I went to in high school which was the Upward Bound program. I enjoyed it and it gave me a lot, so when I was old enough I worked there as a counselor to give back to the program. I really enjoyed working with the kids. While I was at Coe, they did not have a major for secondary education but they did have a program where you could get your teaching license. I left with a chemistry degree and took whatever hours I needed to get my teaching license so I was certified to teach physics and chemistry and I had my coaching certificate.

High school chemistry was very easy for me so I did not change any of my study habits for my first test in college. I ended up getting I think a D on it. I was like, huh, well that's weird. The second test was the first time I actually failed an exam. It was freshman chemistry class and the whole class ended up tanking the test. I remember the professor came in and handed them back and we were all comparing exams and saying how terrible we all did. The professor came in and told us it was our fault, that this wasn't high school and that we needed to change our habits and how we think

https://doi.org/10.1515/9783110798777-005

about this stuff. He walked out of the room, gave himself a couple of minutes, came back in and said okay next chapter. He was an interesting professor and I learned a lot from him. He was my organic professor too, and he was brilliant and great to learn from. It was an awakening moment and I remember it was one of the shocks in college. Also, I remember thinking that I had to study and do the problems in the book. It was a growing up or maturing moment for me.

Upward bound
This is a grant program run by the Department of Education. It "provides fundamental support to participants in their preparation for college entrance." It focuses on getting high school students to graduate and pursue higher education. Eligibility is based on income in addition to children of parents who do not have a BS degree.
https://www2.ed.gov/programs/trioupbound/index.html
08/25/23

5.1.3 Did you do any research as an undergraduate? If so, what was the project?

No I never did other than the typical lab stuff.

5.1.4 Do you have any words of wisdom for students wishing to pursue a chemistry (STEM) degree?

Accept the challenges and maintain your curiosity. It sometimes gets a little tough, but don't lose sight of the trees for the forest is a great analogy. There is a lot of stuff going on that could easily distract you.

5.1.5 How often do you use what you learned as an undergraduate in your current job?

Great question. I would say that the critical thinking and basic chemistry skills I use about half of my day maybe even more than that. I did take some biology courses while I was there and that helps with my yeast and fermentation work. Certainly on a daily basis.

5.2 Movement to brewing field

5.2.1 What was your "entry beer" into enjoying craft beer? Why did you like it?

I know what you mean. It was an interview question when I first started. I think it was a Michelob Amber Bock. I tried something outside of the standard beer that I had before. There was a lot more maltier flavor to it and it had more character and real flavor. I then started exploring from there. Bock is one of my favorite styles and I actually just made one yesterday.

Michelob Amber Bock
It is hard to find this one nowadays. This 3.8% ABV beer is described as being "brewed using 100-percent malt – including dark-roasted black and caramel malts and all-imported hops – Michelob Amber Bock has a unique, rich amber color and smooth, full-bodied taste."
https://drizly.com/beer/ale/red-ale/michelob-amber-bock/p19878
8/25/23

5.2.2 Did you get started with homebrewing?

Yeah, I actually did. I was gifted a homebrewing kit and I thought well that is cool. I never thought of making my own beer. I made that first batch and wanted to explore it more. I ended up with more carboys and more buckets. Then I went from bottling to kegging and at that point now my homebrew system is a half-barrel system and I have stainless steel tanks.

5.2.3 What was your first recipe?

It was a Cooper's Ale kit if I remember right. I'd have to go back and look at my notebook to make sure. That's why we write stuff down, right?

Cooper's Ale
Cooper's Ale is a 5.8% beer that has been brewed as far back as 1862. It is sold as a bottle-conditioned beer making it rather unique as a commercial offering.
https://coopers.com.au/our-beer/ales-stout/sparkling-ale
08/25/23

5.2.4 Did you ever enter a recipe into a brewing contest?

I did a couple of years ago. It did alright. The problems that it had going in I knew that it would be an issue coming out in the judging. It was not one of my best batches. It was not my best beer, but it was what I had the time to send in. My goal was just to enter a competition. I think I got a 31 or 32. It had some flaws and it did okay. I am hoping to enter some more to competitions to help me make better beer.

BJCP score sheet
If you're wondering how a beer score is generated look no further than the judging sheet. Judging categories include aroma, appearance, flavor, mouthfeel and overall impression. What I tell new judges is that it is helpful to break up each category in terms of grades (what else would you expect from a professor, right?). Aroma, for example, has 12 available points. So if you break that up an "A" would be 10–12 points, a B would be 7–9 points, etc.
https://www.bjcp.org/wp-content/uploads/2011/11/SCP_BeerScoreSheet.pdf
08/25/23

5.2.5 What was your biggest disaster?

When I started to upgrade to all-grain I had a brew buddy who was moving out of state and gave me a 15-gallon kettle for some reason. He had only used it once and did not want to take it with him. He made me an offer that I could not refuse. So, I bought that and a bright tank from him. The first time I used it I threw it on the kitchen stove and turned it on and made my beer. When I was cleaning everything up there was a spot that I just could not get off. So, I put it on the counter next to the stove and it had the arc of a circle and I was wondering if I almost caught the counter top on fire from the heat reflecting off the pot. My wife had enough and told me to move my brewing out into the garage. We live in an old farmhouse so the countertop just has more character now and I didn't get into too much trouble though.

5.2.6 Were you part of a homebrew club? Which one? How long a member?

No. I have never been part of a club. It is something that I've played with on my own and I experiment on my own. I am affiliated with AHA, but not an actual club.

5.2.7 What was your biggest batch as a homebrewer?

The biggest batch that I brew is about 15 gallons. I end up yielding around 13 out of the fermenter due to trub losses and hops.

5.2.8 Did you ever brew a beer for an event (wedding)? How did it go?

No I never did anything for an event. I just throw the kegs in my kegerator with five taps and I just go out and pour myself a beer.

5.2.9 Prior to getting into brewing what else did you do after graduating?

After I graduated I went into the teaching profession. I taught for a year and left the industry for a year. I kind of wanted to see what else was out there. I ended up working in a quality control lab for a chemical company and that was not what I was looking for. I was also missing working with the kids so I got back into education for an additional 9 years or so. I taught at a few schools and enjoyed my time teaching and coaching. I enjoyed my time doing it. I was at a brewery and looking for a change of pace after Covid. I taught for 1 year afterward and it was rough. I started looking for something new. The local brewery happened to be looking for a brewing assistant. I remember that I was sitting at the brewery with my wife 1 day and complaining about work. She told me to hop on the Indeed App and see what was out there. It turned out that the brewery that I was at was looking for an assistant brewer.

5.3 Current position

5.3.1 What are your current job responsibilities and tasks?

My title is the lab technician and cellar lead. I wear two different hats most of the time. I do anything related to the lab and it all has to go through me. There are a couple of people where if I'm sick or on vacation they can take the lead and take care of the day-to-day tasks. The lab side includes daily checks on all the tanks that are actively fermenting and then it involves making decisions on whether the tanks are going to get hop additions, temperature changes and so we know that the tanks are where they are supposed to be. We don't want tanks that are lagging.

I am also in charge of the yeast. Once we pitch the yeast we will serial repitch for several generations. It is my job to make sure that the yeast we are pitching is healthy and ready to be used to make more yeast to make beer. I do a lot of yeast counts and make sure that we have viable yeast and that we are using the right yeast and we make sure that the right strain is going from tank to tank. We want to make sure that we are not putting lager yeast into an ale for example. I do active documentation of yeast flavor so that if we have an off-flavor in a tank we can go all the way back through the process to see where a problem might be.

With the cellar lead, it kind of plays off of the lab tech side of it. I test the tank and the gravity of that tank and either test it myself or have somebody else do it and I pass that information along to someone in the cellar and we make adjustments to the tank so that we can make all the different styles that we are producing. Fruited sours have a very different fermentation process as far as when we take action steps as compared to say our IPAs. Even at that, we treat our Hazy IPAs differently than our American IPAs just because of the difference in the styles and yeast strains. We make sure that our tanks are clean and sanitized. We do ATP testing to make sure we have no microbes in the tank after we clean. We started doing that shortly after I started. It is very important for us to do that. Our clean-in-place (CIP) cycles have gotten pretty good. Almost every time when we finish a tank it is zero and that feels good to have a shiny tank with no microbes in it.

Cellar lead

The exact nature of cellar lead is going to depend on the brewery in question. Job descriptions for the position include packaging, operating canning lines, barrel aging logistics, wort cooling, data entry, bright beer work, equipment maintenance and developing or following standard operating procedures.

5.3.2 What is your favorite and least favorite process to do and why?

Sometimes the cleaning of tanks can get a little old. That is probably my least favorite. We moved another beer today, so we have to clean the tank now. That can get pretty monotonous. I like working in the lab. I have a science background and that is why I like it. Yeast counts are not too bad. I also like going to a tank the day after we brewed and pitched the yeast to see a really nice gravity drop and a pH drop. We know the fermentation is going to be done on time and that is a really nice feeling to know that we could get the fermentation started off right and that keeps our brewing schedule on time. When we have problems and it doesn't start off and that is not fun. I have been there long enough to know what action steps we need to take when that happens to correct some of those things.

5.3.3 What classes that you took apply to some of the things that you have done in the brewing industry? Why/how?

I took a cell molecular biology class just for fun one time and because I needed an elective. It was very helpful for understanding the enzymatic reactions that go on during the process of making the wort into beer. As far as chemistry goes, I do not get to do anything crazy such as looking at spectroscopy or anything. I do serial dilutions that you would get from freshman chemistry.

5.3.4 What adjustments did you need to make from working in a QC lab setting to a brewery setting?

I would say that the biggest difference between the two was that I would just stay in the lab unless I was going out to sign off on a quality control check. I did not even have to go out and get the samples because they would come to me. Where I am now I am always out of the lab getting samples, which is not a terrible thing. I really enjoy the flexibility and like solely focusing on the lab. I may have to go pick grain for the next week's brew day or it could be loading the truck. It is important to be flexible and just to go along with more than just my job description. I could be putting cans in boxes after I get my lab stuff done. The next day I could be loading palettes of those same beer cans onto a semitrailer. It is a need to be flexible enough to wear my lab hat one moment and then loading a truck the next.

5.3.5 What do you enjoy most about your current position?

I would say my team and really just being in the beer industry. We have a tight knit team that is great. I love being part of a team when you are firing on all cylinders. I love the industry and the people in the industry. It is a lot of fun. It is great to be a part of the beer culture.

5.3.6 Do you have any words of wisdom for students wishing to apply their chemistry degree toward getting into the brewing industry?

You might not always get that glamorous lab tech position where you are in a lab with all the fancy equipment. You may have to start out with washing kegs and washing tanks. You may have to grow up into that lab role. The number of breweries around here has grown so much and you just never know where it is going to take you. I went from teaching AP chemistry to being a lab tech. You just never know. I spent a lot of time washing tanks earlier than I have to now. You can get your foot in the door by washing kegs and tanks. Hard work pays off in the industry.

5.4 Experiences in the brewing industry

5.4.1 How would you describe your leadership style?

I like to describe it as more of a "Do as I do." I really don't like to tell people to do something because I said so. It should be more that I want you to watch how I do it and then do it that way type of leadership style.

5.4.2 What are the various ways that you keep yourself grounded and you use to take care of yourself?

Leave work at work. Whatever happened at work try to leave it at work as much as possible. Exercise is another way, like right now I know now that weather is getting better and I'm going to try to get a jog in. Once you walk out at the end of the day then leave it there.

5.4.3 Have you had any negative experiences in the brewing industry that you would like to share?

I don't know if there are any specific things, but I will say that mistakes happen and try not to take them too hard. I could share some generalized things that happened. We don't harvest our yeast after tank additions. I had one person throw something into a tank after it had been harvested and that can sidetrack the entire brew schedule because now we are short yeast and we can't make as much beer for that style. The important thing is to learn from those mistakes, grow from them and try not to repeat them. I've been told that my grandfather had a great saying, "Well you made a mistake. Just don't ever do it again." You need to be able to learn from your mistakes and not repeat them or get too hung up on them.

5.4.4 Do you have any advice that you would give to your younger self before getting into the brewing industry?

Consider getting into the industry sooner.

5.4.5 Describe your toughest day in the brewing industry so far

I really can't think of one that would really stand out in my mind. Other than some-body maybe making a mistake and dwelling on it too much. We have a saying on our team that so long as nobody gets hurt it's a great day. We can still get enjoyment out of beer and the beer isn't hurt. We just move on.

5.4.6 Anything else you'd like to share about brewing in your particular area?

I can't think of any particular challenges. The industry has really blossomed in the last 10 years so there are so many opportunities. Even when I started with Single Speed almost 2 years ago now I know of at least two, possibly three, breweries that have started since. Even one of those new ones was someone who used to work with us. He has been in the industry over 10 years, found some investors and is supposed to open this summer. Another one opened up a year ago in a town just up the road. One of the downsides to our area is because it is in the middle of the heartland is that our beer culture is not as strong as maybe it is on the West or East Coast. There is a lot of consumer education that needs to happen where they don't understand the nu-ances between craft and commercial beer. They don't have that appreciation for the variations on all of the different styles.

5.5 Professional affiliations

5.5.1 What current professional organizations are you affiliated with? Why did you join them?

I mentioned earlier that I am a member of the AHA. It is more of an amateur organi-zation. I joined it before I was in the profession. The other one is ASBC. I joined that one because I thought that would be great for my position with those resources to help with growing labs and methods. There are a lot of research articles that I can get access to. They are a great organization. Their website offers webinars such as Omega giving a presentation on yeast.

5.5.2 What do you enjoy most about going to their events?

I have not attended any yet. Our Iowa Guild does put on an annual seminar where they get all the breweries to buy tickets, we sent our entire team, they have presenters and we learned a lot. One of the breweries will host it and people like hop producers or others in the industry will be there. They offer that as a learning opportunity.

Iowa Brewer's guild
This guild supports beer, cider and mead brewers. Their mission is to "unify craft
brewing professionals, promote Iowa beer, and advocate for the industry's growth."
They host an annual festival in the early part of the summer.
https://www.iowabeer.org/#home
08/25/23

5.5.3 How important is it to be associated with other professional organizations?

It is definitely a good thing to consider. You can look at it as an investment. You get
access to more literature and more information and knowledge. You can look at it as
an investment in your future. When you are just getting into your field you may think
that you learned so much, but if you look back on it in 10 years you may realize that
you didn't know a lot. So you can fill in those gaps.

5.6 Other/miscellaneous

5.6.1 Outside of brewing what other hobbies or interests do you have?

My wife and I do horseback riding and we have a small herd of horses that we like to
work with. Actually, that's what we're going to do right now. We have a small acreage
and we want to get all of that Spring cleaning done.

5.6.2 Is there anything else that you'd like to share about your personal or professional life that you think people might be interested in knowing?

Nothing that I can think of off the top of my head.

6 Eli Lowe

Montana State University, 2017; BS Chemical Engineering, Chemistry Minor

Bridger Brewing, Montana, Head Brewer

Montana State University
https://www.montana.edu/
08/11/23

Bridger Brewing
https://www.bridgerbrewing.com/
08/11/23

6.1 College/degree pursuit

6.1.1 What inspired you to pursue your degree?

I was going through high school and taking those basic courses that everyone went through. When I got to chemistry it lit a spark in me when I was learning how everything worked in the world around us and on that small of a scale. The more I learned and saw how applicable everything was the more I appreciated it – literally everything around us is chemistry. It seemed very interesting and a field with endless knowledge to be gained. There is so much you can learn from chemistry.

6.1.2 How long did it take you to graduate? Were there any stumbling blocks you'd like to share?

It took me 4 years. I definitely had to take a few courses over again. I was fortunate that I still made it out in 4 and had to hustle my last 2 years. I had my fair share of stuff not making sense the first time around. Organic chemistry really kicked my butt the first time around. At the end of the day if it feels like something you want to do, you just need to keep at it and eventually it will click. I was fortunate that my school offered courses in such a way that I could retake them in order to keep up. Physical chemistry was another one and it did make me question whether I wanted to do chemistry anymore or not.

https://doi.org/10.1515/9783110798777-006

6.1.3 Did you do any research as an undergraduate? If so, what was the project?

No, not really except outside of my senior project. On the chemical engineering side I did my capstone on brewing, and that is how I first got interested in the field. It was a year-long project and we were tasked with designing a full brewery system from raw materials to yeast metabolism to a finished project. For my chemistry minor I did a project on metabolics and how certain basic nutrients played an important role in psychological help. More specifically I looked into how zinc and selenium can take part in how postnatal depression takes hold in women in relation to deficiencies or surpluses of those.

Zinc and postpartum depression
Zinc has been suggested to help reduce the length of colds and, interestingly, is an important nutrient for yeast. Several studies indicate that pregnant and postpartum woman may not receive an adequate supply of zinc during or following their pregnancy. The cited study determined that zinc supplementation postpartum may have a positive effect on decreasing postpartum depression.
https://doi.org/10.3390/medicina58060731
08/11/23

6.1.4 Do you have any words of wisdom for students wishing to pursue a chemistry (STEM) degree?

I would say it may be hard, but if it's something that you want to do then it will be worth it eventually. A lot of it can seem tedious and boring and tough to do at times. In college, they do teach theory and once you get into the practical side it will all start to make sense.

6.1.5 How often do you use what you learned as an undergraduate in your current job?

Definitely the basics I use every day. You know, chemistry, biochemistry and thermodynamics. Fluid mechanics would be another one. It applies every day from a process standpoint. The yeast biochemistry and Krebs cycle as well as glycolysis definitely play a role in working with the yeast and getting it to do what you need it to do.

6.2 Movement to brewing field

6.2.1 What was your "entry beer" into enjoying craft beer? Why did you like it?

That would have to be Fat Tire from New Belgium. I like that it was still a light beer, very approachable. If you were, for example, to try an IPA for the first time it might turn you off of beer. Fat Tire is easy to like and it's not a huge jump from the traditional commercial lagers. Also, it is a little sweet and pleasant as opposed to a bitter hop bomb like an IPA.

Fat Tire

Fat Tire is produced by New Belgium Brewing which originated in Fort Collins, Colorado. It is another beer that introduced people to the joys of craft beer. This 5.2% ABV beer "blends a subtle malt presence with a slightly fruity hop profile and crisp finish, to create a bright and balanced beer that drinks easy, anywhere." It is produced with a house ale yeast and the hops used to make it are Triumph, HBC-522 and Barbe Rouge.
https://www.newbelgium.com/beer/fat-tire/
08/11/23

6.2.2 Did you get started with homebrewing?

I actually did not.

6.2.3 Prior to getting into brewing what else did you do after graduating?

I actually went straight out of college into a brewing position. I had worked at Bridger Brewing in the kitchen while attending college. At the time I was graduating they were working on plans for a new production facility. So, I approached them and asked them if they were looking for a new brewer to make this happen to work with them and get trained up while they were getting this facility worked out. It was a combination of having a skill set and being in the right place at the right time.

6.3 Current position

6.3.1 What are your current job responsibilities and tasks?

I am the headbrewer of our distribution facility. I basically oversee production scheduling and recipe scale ups for large-scale distribution. I also oversee our lab manager and verify that everything is okay on the QC/QA side of things. Additionally, I handle basic staffing organization and production scheduling.

6.3.2 What is your favorite and least favorite process to do and why?

Brewing is still my favorite process for sure. It's because I know that is where a lot of the biochemistry happens and I can watch the process move along. There are no unintended issues with the process. It is nice to know when you have a batch of beer that comes out how you wanted it to and that it came out the way it did in the prior batch that you brewed. My least favorite is probably equipment maintenance. Stuff wears out over time and breaks down. Then you have to halt production and have to fix it. It is the less glamorous side of brewing but it is still very necessary.

6.3.3 What classes that you took apply to some of the things that you have done in the brewing industry? Why/how?

Definitely biochemistry is a big one. That applies to yeast health and enzyme reactions within the beer. Those basic chemistry courses, knowing how chemicals for cleaning agents work and mixing in those concentrations. Obviously knowing not to mix certain chemicals. On the lab side, having those lab skills from college that translated to the QA side of things as you use a mass spectrometer, gas chromatograph and incubators for microbial samples. Those are the biggest translators for working in the brewing industry.

Yeast vitality and viability
Yeast viability and vitality are two often confused terms in brewing circles. Viability refers to the percentage or number of cells that are alive versus the number of cells that are dead. Vitality, however, describes how well the cells are doing. The way that I once explained it to my students is I could use attendance to measure the percentage of students that make it to my class on a given day (viability), but the percentage of students who are actually learning something from me at a given time is lower than that (vitality) – hopefully not too much lower though.

6.3.4 What adjustments did you need to make from working in a lab setting to a brewery setting?

You have to be more creative with your setup. An academic lab will have more of the bells and whistles associated with it than a brewery-based lab. A lot of times you have other things that you have to work with on a limited base such as glassware. There is a longer turnaround on your tests because you have glassware that you need to clean before you can run another test. You can't just run them all at once. It can constrain you with the access to materials and the like.

6.3.5 What do you enjoy most about your current position?

I enjoy the flexibility of the position. For many of the times that we are caught up on production and preventative maintenance then we can go home and enjoy what is left of the day. So long as there is something that is not directly tied to a scheduled task, people can come in after a doctor's appointment in the morning and it isn't a problem. Everything moves in a way that flows, and so long as it gets done within a relatively quick amount of time it is easier to navigate. Once the beer is brewed you have 2 weeks until it's finished so it doesn't matter when you really start. You could get started at 8 AM or 8 PM and it does not really matter that the 12 h will make a difference in the long run.

6.4 Experiences in the brewing industry

6.4.1 How would you describe your leadership style?

I try to be more of a guide than a manager. I am responsible for a lot of the training of personnel in the facility. We have some hazardous materials training and all of that which goes on with it. I like to be a resource for people and not be overbearing. I'm a fan of the lowest level of decision-making and giving people the freedom to do that. If there is a question that comes up, I am more than happy to help with it. I have this huge fear of being a micromanager and don't want to be overbearing. I want to give my employees space but also want to be helpful to them.

6.4.2 What are some unique strengths that you think chemical engineers bring to the brewing industry?

More of the physics-based background. Learning more of those thermodynamic principles and fluid mechanics as well as heat transfer applications. I did not see as much of that in the straight chemistry classes. I would also say more of an understanding of the larger process scale as opposed to the minutiae of small-scale chemistry of what is going on inside the tank.

6.4.3 What are the various ways that you keep yourself grounded and you use to take care of yourself?

I try to not let this job consume me as it were. You need to have a passion for brewing in order to do it and do it right. It is very much a labor of love in a lot of ways. I try not to overwork myself. If I have to take a break then I will and I try to encourage everyone else to have that same mindset. You need to be able to tell people to take a break and not be here all the time. In doing that, it helps me to accept that as well.

6.4.4 Have you had any negative experiences in the brewing industry that you would like to share?

There can be quite a bit of crunch. For example, when we were just getting started up here there were times when I was working 80–100 h a week during start-up. You know if something goes down on the production side and you are behind on production you can lose a lot of time and it will take you some time to catch up. For every hour that you lose it will take you 3–4 h to get caught back up. That is probably the biggest downside to the industry.

6.4.5 Do you have any advice that you would give to your younger self before getting into the brewing industry?

I would have gone more into biology and biochemistry in college. It was a weak spot of mine coming into the industry. It was not as much of an emphasis during my undergraduate degree. So things like yeast health and metabolism took some catching up for me to do.

6.4.6 Describe your toughest day in the brewing industry so far

When I did my first solo brew I really botched it. I ended up with a stuck mash. It is where you can't get your wort through the grain for brewing. It took me 7 h to get enough wort out to be able to boil with. So what was normally an 8-h brew day was already hitting 10 h and I had not even started boiling or anything. On top of that the yield was bad. There were quite a bit of rookie mistakes. It was a wasted batch and it made me question whether I was actually cut out for this. My bosses were really understanding and told me that it happens to everyone and we all have bad days in brewing. For me, it was just to have it happen so early on and you have to live and grow with it and learn.

Stuck mash

One of the major culprits for extending the time associated with a brew day is the dreaded stuck mash. This occurs when the brewer cannot get the water to adequately drain from the mashed grains during sparging. A number of factors can affect stuck mashes including whether the grains have been milled too fine, characteristics of the grains themselves or if the water to grain ratio was incorrect during mashing. One method that brewers use to avoid stuck mashes is to add rice hulls to the milled grains prior to starting to mash. Adding a mash out step where water temperature is raised to 170 °F prior to sparging can also be helpful.

https://www.homebrewersassociation.org/how-to-brew/5-tips-for-avoiding-a-stuck-sparge/

08/11/23

6.4.7 Anything else you'd like to share about brewing in Montana or your area?

Our liquor laws are a little different. We have some current legislation that might change that before this book is published. Right now, breweries that have taprooms cannot serve more than 48 oz of beer to a person per day. That is a big sticking point. Also, you cannot share licenses. So, if you have two facilities like we do you cannot share your license. In Montana right now beer that is sold in your taproom has to be brewed onsite.

State liquor laws

Anyone who has ever been trained to serve alcoholic beverages understands the subtleties associated with adhering to state liquor laws. The referenced article provides an interesting window on some of the more unique laws. At the time of publishing, only four states close their liquor stores on Sunday while eight states do not allow "happy hours." Nevada does not have a "last call" and, in fact, allows alcohol to be purchased at any time.

https://www.usatoday.com/story/money/2023/06/25/alcohol-laws-by-state-us/70344352007/

08/11/23

6.5 Professional affiliations

6.5.1 What current professional organizations are you affiliated with? Why did you join them?

I am affiliated with the Montana Brewer's Association and the Brewer's Association as a whole. They are the best resources in the field. Like for beer trends or how the industry is as a whole. For example, if there are emerging beverages such as seltzers or nonalcoholic beverages you learn more about those. ASBC has also been helpful for learning the procedures and testing methods for our beers. The final one is the Master Brewers for professional brewing techniques.

6.5.2 What do you enjoy most about going to their events?

I've only been able to attend GABF in the past so far. I really enjoyed it because it was nice to see breweries from across the whole nation represented and kind of be this great coming together for breweries. There is a lot of variability in how people do things and learn about new ways to brew and styles.

6.5.3 How important is it to be associated with other professional organizations?

I think it can be helpful from a resource standpoint to know where to go if you have a question or need to know how to do your job better.

6.6 Other/miscellaneous

6.6.1 Outside of brewing what other hobbies or interests do you have?

Not to play too much into the nerd stereotype, I like Dungeons and Dragons. I also like to camp and fish and hike and be outside. A lot of times I Iike relaxing and being able to turn it all off.

6.6.2 Is there anything else that you'd like to share about your personal or professional life that you think people might be interested in knowing?

Don't think so. We pretty much covered it.

7 John Paul Maye

BS, Chemistry, Minor in Biology, Northeastern University, 1988

PhD, Organic Chemistry, Purdue University, 1994

Hopsteiner, Technical Director

Northeastern University	Purdue	Hopsteiner
https://www.northeastern.edu/	https://www.purdue.edu/	https://www.hopsteiner.com/
08/07/23	08/07/23	08/07/23

7.1 College/degree pursuit

7.1.1 What inspired you to pursue a chemistry (STEM) degree?

I actually started out as a biology major in college. I attended Northeastern University in Boston. They had a very advanced Co-op program. After my first semester I met with my Co-op advisor and when I told him I was a biology major he told me that I might have to work for free because there weren't too many well-paying jobs for biology majors at the time. Then I said "well I'm thinking of minoring in Chemistry" and he said "oh well there are lots of great paying jobs for Chemistry majors" so I ended up switching to chemistry and I got my minor in biology. In fact, my first Co-op job was with Uniroyal Chemical in Naugatuck, Connecticut in the Polymer Research group and I ended up Co-oping with them three times over the course of 15 months and then I had another Co-op job with Aurther D. Little in Cambridge, MA for 3 months and did critical propane extraction work with them. My last Co-op job was with Shipley, in Newton, MA that made a chemical called photoresist which is used in the electronics industry. I realized through my Co-op jobs that the real interesting work took place at the PhD level. I applied to a number of universities for graduate school and ended up at Purdue University working under Dr. Negishi and researched zirconium chemistry.

https://doi.org/10.1515/9783110798777-007

Co-ops versus internships
While both of these involve working at a company while pursuing a degree the major
difference is in the number of hours required. Co-op students may work close to full-time
hours (e.g., 35–40) per week whereas Internship students often work for fewer hours.
https://www.usnews.com/education/best-colleges/articles/2015/03/31/understand-the-dif
ferences-between-a-co-op-internship
08/07/23

7.1.2 How long did it take you to graduate? Were there any stumbling blocks you'd like to share?

Because of the Co-op program it took me 5 years to graduate from Northeastern. What
I like to point out is because I was both a chemistry major and biology minor I had
lots of labs worth one credit each and also I took a two credit scuba class. When I
started my last year I added up all my credits and I realized I have the exact number
of credits to graduate at the end of my first semester senior year. I did not have to
take any classes that last semester. I was afraid the school would come back and say
that I was off by one credit but it never happened. So it ended up working out nice for
me; I did a Co-op during my last semester.

I actually did drop one class and it was a Spanish class. When I told the teacher
that I wanted to drop her class she asked why because I had a solid B. I said I know
but I'm taking physical chemistry, a biology class and a physics class and all of them
have labs. And p-chem (physical chemistry) was one of my hardest classes and I just
don't have the time for Spanish. I'm spending a lot of time studying Spanish compared
to my other courses. I took three p-chem courses as an undergraduate.

7.1.3 Did you do any research as an undergraduate? If so, what was the project?

I had an opportunity but it wasn't well organized back then. They were probably just
trying to give me some exposure to work back then. I got more exposure through the
Co-op jobs that I had.

7.1.4 Do you have any words of wisdom for students wishing to pursue a chemistry (STEM) degree?

Try to do a Co-op program. I made really good connections with companies that I
worked for. When I told them I was heading to Purdue for my PhD both Uniroyal and
Shipley told me if graduate school doesn't work out you have a job here. They had
already seen my work ethic and they knew I was a decent guy. That was actually a

nice benefit because when I went to graduate school it seemed like there were two types of people that I met there. What I mean by that is the ones that found their undergraduate very easy found graduate school very hard and vice versa. I found my undergraduate to be very hard but graduate school was easier. Again, I think the other part of it too is I remember it was nice to know that I had two job offers heading into it. Others at graduate school were very envious that I had this work experience. I always tell students even during the summer if you can get a job so you can get some work experience it is worth a lot.

7.1.5 How often do you use what you learned as an undergraduate in your current job?

That's a bit of a harder one because I'm a technical director. I think I really became a real scientist after graduate school. The basic chemistry I learned was there and can be applied to a bench chemist's job. With a PhD you become a researcher.

7.2 Movement to brewing field

7.2.1 What was your "entry beer" into enjoying craft beer?

When I was going to school in Boston, Sam Adams was a new kind of beer that was being produced and sold there. Then eventually it was Sierra Nevada and New Belgium. In the early days of working in the hop industry when I attended brewing conferences the beers served were the same three to four beers every time. Now today when you go to those conferences there are so many craft brewers attending and there are dozens of different craft beers to try.

Samuel Adams Boston Lager
According to the company website, Samuel Adams is a 5.0% ABV full-flavored beer that was first made in 1984. It underwent a redesign recently and is being marketed as being lighter in color. The beer is described as "being complex and balanced in malt and hop flavors."
https://samadamsbostonbrewery.com/
08/07/23

7.2.2 Did you get started with homebrewing?

I did. That was one of the reasons why I applied to Pfizer. I still brew today.

7.2.3 What was your first recipe?

It was probably an ale and most likely the malt extract type of recipe. I didn't have the equipment for all grain then.

7.2.4 Did you ever enter a recipe into a brewing contest?

I have not, but people ask if I could make some for them or sell it to them. I keg now but I used to bottle.

7.2.5 What was your biggest disaster?

I made a beer I called MA for more alcohol. I brought up a case to Northeastern and there was one night when I heard this explosion and some foaming. A couple of bottles exploded and so I think I ended up putting them in the refrigerator to keep them cold – wanted to stop re-fermentation. It was highly carbonated more like a champagne beer.

7.2.6 Were you part of a homebrew club? Which one? How long a member?

I haven't gotten any notices from them recently and have not done a lot with them. What I find sometimes when I meet with some homebrewers they are intimidated by my technical background.

7.2.7 What was your biggest batch as a homebrewer?

Just the five gallon batches.

7.2.8 Did you ever brew a beer for an event (wedding)? How did it go?

No. I do usually throw an Oktoberfest party or St Patrick's Day party and we'll drink that beer that I made.

7.2.9 Prior to getting into brewing what else did you do after graduating?

I went straight from school to work.

7.2.10 Would you recommend brewing to other chemists out there and why?

It's a fun industry to work in. I would say to try to get a job at one of the larger breweries because they can pay you more. I'm kind of surprised by the lack of lab space that some of these smaller ones have. The number of labs is getting fewer due to some of the consolidation that has occurred in the industry.

7.3 Current position

7.3.1 How did you end up in the hops industry?

Back in those days we didn't have the internet so I went to the library to look for jobs and ended up coming across a book that had every chemical and pharmaceutical company in the country. Anyways I'm from Connecticut originally and knew that Pfizer had a division in Connecticut. The directory listed all the locations that Pfizer had throughout the country. They had a listing for their brewery and dairy ingredients division in Milwaukee, Wisconsin. I was like wow, I make beer as a hobby and I mailed them my resume and it so happened that the guy who was running their brewery group was looking for somebody and they called me up and said "we got your resume, we're hiring and we are looking for somebody with your background. Could you come up here for an interview?" So I drove up from West Lafayette in Indiana to Milwaukee and I interviewed and did a good job with it. They asked me back for another interview and so I went back up and met some other people and they offered me the job. The Pfizer brewery division was doing research in hops and they were selling these specialty hop products to the brewing industry. An interesting side note is how Pfizer, a pharmaceutical company, ended up in the hops field? It turns out that beer can get skunky when it gets packaged in clear or green bottles and that skunkiness was caused by hops, iso-alpha acids. That discovery was made by a couple of researchers at Asahi back in the 1960s. When research chemists at Miller brewing learned about it they found that if you treat the iso-alpha acids with sodium borohydride one can reduce the side chain and make the hop acid light stable. They were able to demonstrate that and they got a whole bunch of patents on that research and technology. Pfizer bought a company in Milwaukee called Paul Lewis Laboratories that sold enzymes to the dairy industry for making cheese. They were also selling enzymes to the brewing industry. Papain, for example, is used to clarify beer. They wanted the dairy business but they ended up with the brewery business as well.

In regard to the dairy industry there is an enzyme in the intestinal tract of cows that Paul Lewis was extracting – when it's added to milk it turns into cheese. Pfizer had developed a bacterial process to make this enzyme so that it was vegetarian-approved. So, they wanted to get into the business and the only way to do that was to purchase a company that was producing the natural enzyme from the intestinal tract. They ended up inheriting the brewing business. Miller brewing company is not a chemical company so they reached out to Pfizer's brewery and dairy ingredients division in Milwaukee. When Miller made this sodium borohydride discovery they contacted Pfizer and said since you're working in the brewing industry we have this chemical process so do you think you could make this product for us. In order for them to make the desired product they had to use hops extract and in those days hops were being extracted with organic solvents such as hexane. Pfizer purchased a hexane extraction plant in Sydney NE, bought hops from Yakima Washington, and would take tanker cars full of this hexane hop extract and ship it to Groton, Connecticut. Then they would do the sodium borohydride treatment and fish out the resulting light stable rho-iso-alpha acids product and formulate it into a 30% solution and sell it to the brewing industry. I got to Pfizer a few years after that as they were developing several new products from hop extracts. They were also able to hydrogenate the iso–alpha acids with palladium on carbon and reduce the double bonds on the iso-alpha acid molecule to make tetrahydro-iso-alpha acid and what's neat about this is you can add very small amounts (4–5 ppm) to beer and get fantastic foam and lacing. That product was relatively new when I joined so my job was to go around and give technical presentations to breweries around the world on these products and their various benefits.

7.3.2 What are your current job responsibilities and tasks?

A lot of what I do is try to develop new hop products from hop extract and find new applications for those hop products outside the brewing industry. Hops are known to contribute bitterness, antimicrobial protection as well as antioxidant protection to beer. We know the compounds in hops that contribute to those characteristics. The alpha and beta acids are very antimicrobial. In fact, the beta acids can inhibit almost all Gram-positive bacteria at a 1 ppm dose rate. It's almost as effective as some antibiotics. One of the challenges is that beta acids are not very soluble. So, it really depends on the type of application. Alpha acids are also antimicrobial but they have a tendency to isomerize to iso-alpha acids, but iso-alpha acids are also very antimicrobial around 8 ppm. One of the bigger accomplishments that I had in this field was that a lot of fuel ethanol plants were using antibiotics to prevent bacteria growth and so I was able to show them that they could use our 30% iso-alpha acid product as an antimicrobial. Some of those hop products are also used in the yeast industry for yeast propagation. Any kind of fermentation is going to face a bacterial challenge, primarily

Gram-positive bacteria because a lot of them are anaerobic. Hop acids work very well against those.

7.3.3 What classes that you took apply to some of the things that you have done in the brewing industry? Why/how?

Primarily organic chemistry and some analytical chemistry. Almost all of my counterparts in the hops industry were analytical chemists. Everyone was trying to measure how much alpha acids were in the hops or how much linalool is in hop oil. I was actually one of the only organic chemists and so I could look at a molecule and figure out what else you could do to it and the analytical chemists didn't look at it that way. I think the other thing that helped me out was being a homebrewer. I had a real interest in the application. So, I'd often make comments to people at the company about us making these different products. Those different products contributed a different kind of bitterness so we can market them differently. I started steering people toward those differences in the bitterness profiles. In fact, people would come up looking for that "guy" who made the diagram of the tongue showing where the different hop bitterness profiles interact with different parts of the tongue – me (see Figure 7.1). It's

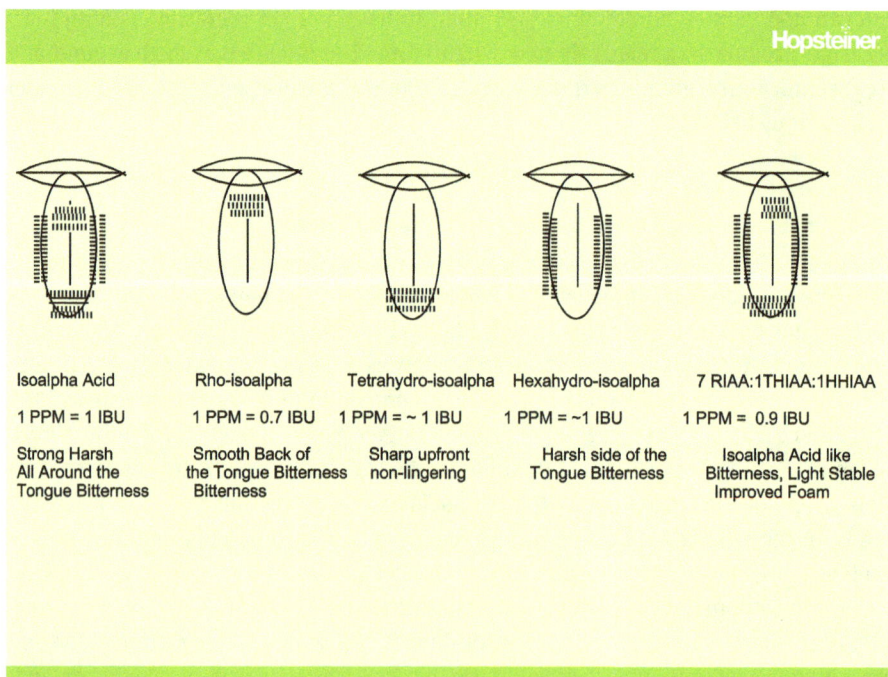

Isoalpha Acid	Rho-isoalpha	Tetrahydro-isoalpha	Hexahydro-isoalpha	7 RIAA:1THIAA:1HHIAA
1 PPM = 1 IBU	1 PPM = 0.7 IBU	1 PPM = ~ 1 IBU	1 PPM = ~1 IBU	1 PPM = 0.9 IBU
Strong Harsh All Around the Tongue Bitterness	Smooth Back of the Tongue Bitterness Bitterness	Sharp upfront non-lingering	Harsh side of the Tongue Bitterness	Isoalpha Acid like Bitterness, Light Stable Improved Foam

Figure 7.1: Courtesy of John Paul Maye, PhD.

funny and interesting you know how people always say that bitterness is just in one part of the tongue but you have it on the front, the side and here and there. I ran into someone who was a sensory expert and who told me that you have bitterness sensory all over your tongue as well as the mouth. The other thing too, we had a customer once who said his beer had too harsh of a bitterness to it. So we said, "why don't you remove some of the hops from the kettle and replace it with the rho product because it has a nice smooth bitterness." So he did that and then he said he was missing aroma and so I told him that we have this other fraction from the CO_2 extraction called beta aroma extract and it contains the hop oils. It contains the aroma, but no bitterness. He started making that beer and then he liked that the tetra product could enhance the foam and lacing of his beer so he ended up removing more hops from the kettle. Long story short, he ended up changing his beer into a light stable beer even though it was packaged in a brown bottle. He liked the fact that he could add the same amount of product each time.

7.3.4 What adjustments did you need to make from working in an academic lab setting to an industry setting?

Because I work in the hop industry I went from working in a lab in an academic lab to working in a lab in the industry. I did have two and then three people that I had to manage. So, I was also managing that and directing their activities. I converted more to working as a manager.

7.3.5 What do you enjoy most about your current position?

I've been pretty lucky actually. They've allowed my curiosity to move things forward. In fact, there's a compound in hops called humulinones and it looks like an iso-alpha acid molecule but it has an extra hydroxyl group. Anyhow, the guy who was before me, he ended up doing a poster on humulinones and hops. We do a lot of beer analysis in our labs and one of the guys in our lab made the comment that if you look at this beer it has 15 ppm humulinones in it. I go, that is really high for those. So I break out a piece of paper; 0.2% of humulinones are in hops, right? If these craft brewers are dry hopping at 2 pounds per barrel and 100% of the humulinones get into the beer that is like 15 ppm so they have to be really soluble. So we started doing some dry hopping experiments and then we discovered all kinds of crazy stuff that was taking place. When brewers dry hop they think they are adding essential oils to the beer and they are. What they didn't know though was that the leaf material can absorb and remove the iso-alpha acids in the beer and then the hops from the dry hopping can add all these humulinones. Now those are only about 66% as bitter as the iso-alpha acids. The other thing that happens is you can also incorporate alpha acids into the beer and those are about 1/10 as bitter. What was happening is that a lot of these craft brewers would do the IBU analysis, an iso-octane extraction on their beer to measure the iso-alpha acid concentration. All of the hop acids in the iso-octane are supposed to be iso-alpha acids but what was happening was when they dry hopped the IBUs would go up, but the beer would taste less bitter. What was happening was the leaf material was absorbing the iso-alpha acids and replacing it with the humulinones and alpha acids which are less bitter. And humulinones and alpha acids also absorb at the same wavelength as the iso-alpha acids. I gave this presentation showing what was happening and everyone looked at me and they were excited because I explained what was going on. I've probably given that talk a dozen times and we have four publications from that work.

Then the hazy IPAs came out and we had just finished all that work on dry hopping. We did a lot of testing on those too and it turns out that the haze acts as an emulsifier and it solubilizes a lot of nonpolar compounds that would not be found in beer or at little concentrations. One of the crazier things is that beta acids are generally not found in beer because they are not soluble. In the hazy beers though we found that the average beta content was 4 ppm and we had some hazy ones that had up to 14 ppm beta acids. It also elevated the alpha acid concentration. We did a publication on that research as well.

One of my biggest accomplishments I have to say within the last year is that most people are familiar with the three major homologues of alpha and beta acids. There are also three minor homologues that are found at a low concentration. We found a fourth minor homologue recently and I just presented at the EBC meeting during June 2022 and we submitted a paper for publication in the *ASBC Journal* that was published on June 23, 2022 (https://doi.org/10.1080/03610470.2022.2079944).

7.3.6 Do you have any suggestions or words of wisdom for students wishing to apply their chemistry (STEM) degree toward getting into the brewing industry?

Just getting into the brewing industry is your first step. I think what will end up happening is there will be some opportunities at a brewery to try different things. You know there may be a chance as a research brewer. There are quite a few people who I work with that were chemists, started working at breweries and are now hop salesmen. Keep your mind open to new opportunities and you may be using your chemistry degree or even end up working on something else.

7.4 Hops industry

7.4.1 How has the hops industry changed since you first started in it?

There has been a huge consolidation. There used to be 100 hop companies and probably four of them today sell 80% of the hops on an annual basis.

7.4.2 What is the biggest challenge to the hops industry right now?

Our focus is now on developing hop products that are more environmentally sound. So, for example, hops are sensitive to powdery and downy mildew and so then you use genetic technology to assist your breeding program. That is we're using the DNA technology to see if the off-spring plants contain the genes for powdery mildew resistance, for example. One of the newer areas that we're working on is crossing hop plants to hops that are known to grow well in a very dry climate. If we can incorporate that drought resistance then we can grow hop plants that need less water.

Downy mildew
This is a disease that is specific to hop plants. It is caused by *Seudoperonospora humuli* and can persist in hop crowns as well as the soil. Wet and foggy weather promote the existence of the disease. Several varieties of hops are known to be resistant including Fuggles and Tettnang.
https://pnwhandbooks.org/plantdisease/host-disease/hop-humulus-lupulus-downy-mildew;
08/07/23

7.4.3 Can you describe your proudest contribution to the hops industry?

Discovering the new hop acid was a big one. All that work we did on dry hopping and the bitterness and the IBU test. That was a really big accomplishment.

7.5 Professional affiliations

7.5.1 What current professional organizations are you affiliated with? Why did you join them?

I'm a member of the ASBC and that is a scientific brewing organization. What is neat about it is people who are an expert on malts are giving talks on malts and those who are experts on hops are giving talks on hops as well as yeast. And others give talks on different equipment and items that help with beer production. I'm also a member of the MBAA. It is more of an applications organization. I've given talks on dry hopping there for example. I'm also secretary of the International Hops Standards Committee. I gave a talk at an ASBC conference in 1998 on making HPLC standards for a number of isomerized and reduced iso-alpha acids. Following my talk there was a hop technical round session where they had leaders of the hops industry at the table. Brewers in the audience were asking questions and a hop researcher said that we need the hop industry to produce those calibration standards that John Paul talked about. A lot of those compounds are being measured using no standards or UV. HPLC is more accurate. Right there they established the Hops Standards Committee and the goal of the committee was to produce these standards which we ended up doing. That committee meets every year and we continually update the standards. If we're running low we make more and so we also add to those. The four standards were produced and then we started making others. Right now the industry needs a de-oiled extract as an alpha and beta acid HPLC calibration standard material. I think we'll make a straight standard after that runs out. The current standard is an extract. You have to warm it up and mix it up or you have solubility issues. If you don't do that the beta acids will surface to the top of the extract. We've had a couple of customers say that they had issues with concentrations and we have to ask them whether they warmed and mixed it up first.

7.5.2 What do you enjoy most about going to their events?

The hospitality suite. Also, meeting with people who you only see once a year.

7.5.3 How important is it to be associated with other professional organizations?

Yeah, if you can do that then that is where you can meet people from other organizations. A lot of companies are starting to cut back on travel for their employees. Some are getting better at getting their employees to attend though.

7.6 Other/miscellaneous

7.6.1 Outside of brewing what other hobbies or interests do you have?

I also make wine as a hobby. I like making bread. You know, fermentation-type things. I also like to hike and I do a lot of hiking. I exercise quite a bit too. It helps keep the weight down.

7.6.2 Is there anything else that you'd like to share about your personal or professional life that you think people might be interested in knowing?

I think homebrewing makes me better at my job. I have a better insight into what some of our customers actually need. Sometimes you don't know what specific question to ask. A customer asks me to send them a sample of something and I ask them if we can talk over the phone to see what their actual issue is. For example they might have heard that beta acids are very antimicrobial and I ask where they plan to use it. They then say in the corn mash and I tell them that it would be an issue because the beta would stick to the corn and not work. Then I would tell them that our iso-alpha acids would be the ones you want to use there. So, that hobby can make me better at my job.

8 Patrick Ting

BS Chemistry, National Taiwan Normal University

MS Analytical Chemistry, Marquette University

PhD Organic Chemistry, University of North Texas

Postdoc: Studied biological oxidation at South Western Medical School, University of Texas

Miller Brewing, Retired

National Taiwan Normal University
https://en.ntnu.edu.tw/
08/10/23

Marquette University
https://www.marquette.edu/
08/10/23

UNT
https://www.unt.edu/
08/10/23

UT Southwestern
Postdoctorate
https://www.utsouthwestern.edu/
education/medical-school/
08/10/23

Molson Coors (Acquired Miller Brewing in 1996)
https://www.molsoncoors.com/
08/10/23

8.1 College/degree pursuit

8.1.1 What inspired you to pursue a chemistry (STEM) degree?

It seemed that my fate was determined by my college entrance exam scores (similar to SAT) which met the rank of one college and major in my wish list (science and engineering) back in Taiwan. You had no choice at that point in time.

https://doi.org/10.1515/9783110798777-008

8.1.2 How long did it take you to graduate? Were there any stumbling blocks you'd like to share?

It took me 4 years for my undergraduate degree, 2 years for my masters, 3 years for my PhD, and 1 year for my Postdoc. Maybe the first year of my master's degree was a stumbling bloc because I had just come from Taiwan so the language and concept of English were difficult for me. I overcame that obviously. I just followed the step-to-step and school by school whatever took me to gain breadth in chemistry and skills as well as becoming accustomed to American life.

8.1.3 Did you do any research as an undergraduate? If so, what was the project?

No I didn't. I did not have that option.

8.1.4 Do you have any words of wisdom or suggestions for students wishing to pursue a chemistry (STEM) degree?

Yes. Our undergraduate program required courses for chemistry majors that covered not just core courses but also many mathematics and physics courses. It was hard and we had no research or internship opportunities available at that time. My first suggestion is that students should try to get some type of opportunity to do research or to work for a company as an intern to learn how chemistry applies to the industry. Right now, I'm retired so I could rethink about all of the chemistry I learned that I used in the brewing industry. I wish that somebody had told me about the real opportunities to do that in my undergraduate studies. I can give you a lot of examples from my work. General chemistry, organic chemistry, physical chemistry. Everything I learned all applied to the research and the application of what I did in my job.

8.1.5 How often do you use what you learned as an undergraduate in your current or former job in the industry?

In the brewing industry as well as others you can't move forward into the future unless you can connect to the past. For example, during my master's degree I was working on GC chromatography and had a summer job to study and analyze cheese flavor in various cheese products for a food company. During my PhD I focused on the topic of organic intermediates for synthesis and application, and my Postdoctorate work was on bio-oxidation reactions using HPLC and an electron spin resonance (ESR) in aqueous systems. All of those techniques that I learned had great impacts on my work and career in the brewing industry and the field of hops chemistry. Hops are one of the indispensable ingredients in brewing. Hop compounds are natural organic compounds that contribute bitterness, flavor, foam, antimicrobial and flavor stability to beer. I studied their chemistry and applications that are involved including the concepts, theories, skills and knowledge bases of chemistry, organic chemistry, physical chemistry and biochemistry. For example, hop bittering compounds are nonvolatile and weak organic acids with a chromophore group which are acid-base interchangeable; in other words, organic chemistry can be performed in an aqueous system (environmental and food safe) and monitored using HPLC and UV. The quality of beer foam is an interaction between hop acids and proteins related to physical chemistry. However, hop essential oils that are responsible for aroma and flavor in beer consist of many volatile compounds which can be analyzed and studied by GC. Prevention of oxidation and biological oxidation is also a constant precaution associated with keeping beer fresh as well as extending shelf-life. You can see how my past chemistry and experience connected to my actual career in the hops industry.

Electron spin resonance

Ready for some quantum chemistry? ESR also goes by electron paramagnetic resonance spectroscopy (EPR) and electron magnetic resonance (EMR). ESR "detects the transitions induced by electromagnetic radiation between the energy levels of electron spins in the presence of a static magnetic field." In this regard, ESR can be used to monitor oxidation-reduction reactions and, in particular, the actions of the enzymes that catalyze those reactions.

https://www.sciencedirect.com/topics/medicine-and-dentistry/electron-spin-resonance
08/10/23

8.2 Movement to brewing field

8.2.1 What was your "entry beer" into enjoying beer or craft beer? Why did you like it?

The first beer was Miller High Life because at that time there was a popular commercial that said "if you have got time we have got beer." It was in a clear bottle that I first tried when I was in graduate school. Also, because I was not a beer drinker, I would put ice in my Miller High Life to make a light beer. I liked it because it would quench my thirst. I was in Texas and it was very hot.

Miller High Life

Miller High Life with an ABV of 4.6% has been advertised as "The Champagne of Beers" for many years now. It has been brewed since 1903 and is "brewed with a proprietary blend of malted barley, hops from the Pacific Northwest, and Miller yeast."

https://www.molsoncoors.com/brands/our-brands
08/10/23

8.2.2 Did you get started with homebrewing?

No, my homebrew started after retiring from MillerCoors.

8.2.3 What was your first recipe?

One of my research projects was to set up a protocol for selecting new varieties of experimental hops. Each variety was single hops at a target of 30–45 IBU and added during the kettle boil for 45 min. The hopped wort was fermented under lagering conditions. In that time, I asked a producer to find an aroma variety that would also be in high alpha-acids content so that I could utilize separately the alpha-acids for making advanced bittering products and remaining vegetative portion for flavoring the beer. I received the special experimental hops and brewed a typical 20 L lager at a target of 45 IBU and aged it for 30 days in our small pilot plant. The beer turned out to be very smooth and rounded in taste and rich in unique and exotic aroma and flavor as perceived by grapefruit, passion, citrus and tropical fruits notes. This was very different from the traditional beers while I was working for Miller Coors. We therefore decided to promote this hop to commercialization and later named it the "Citra" variety. A scale up of a single Citra Hop IPA was later made at Miller Valley brewery (20 BBL pilot brewery) and tapped at the company's bar to advertise the new style of beer. The rise of craft breweries resulted in the Citra hops becoming so popular and thus changed the palates of consumers and the brewing industry. So that is the whole story of the Citra hops.

Citra Hops
Bet you did not know the story of how these were discovered. Yakima Chief, based in Washington, released Citra hops in 2007. This popular hop has an alpha acid content between 10% and 16% and exhibits a lower percentage of co-humulone. The aroma profile for citra hops included citrus, stone fruit and tropical notes. The beta acid content is between 3.0% and 4.5% while the total oil content is between 1.0% and 3.0%.
https://www.yakimachief.com/citra.html
08/10/23

8.2.4 Did you ever brew a beer for an event (wedding)? How did it go?

No. Just to enjoy at home.

8.2.5 Did you ever enter a recipe into a brewing contest?

No. I'm not at that level yet.

8.2.6 What was your biggest batch as a homebrewer?

About 5 gallons with my own home-grown hops. I also grow a few hills of Citra and Cascade in my backyard.

8.2.7 Prior to getting into brewing what else did you do after graduating?

I taught middle school for 2 years and served in the military (ROTC) for 1 year prior to coming to the US after I received my undergraduate degree. Near the end of my post-doc research, I started looking for a job and fortunately found a job in Miller brewing company because of my experience with that summer job studying cheese not to mention my organic chemistry background.

8.2.8 Would you recommend the brewing industry to other chemists out there and why?

I would say yes because there are many opportunities requiring essential chemistry to work as brewers, technical services, quality control or research in the areas of microbiological and bacterial, fermentation and yeast applications, and flavor chemistry.

8.3 Position before retiring

8.3.1 What were your job responsibilities and tasks?

When I was hired by Miller in 1978, the Research and Development department was just at a building phase or initial phase. In other words, there was no specific direction or projects but the only goal was just trying to make beer better. The aspects of my job or responsibility were bitterness, aroma and flavor, and I therefore relied on my own research. One of Miller's major brand, Miller High Life, was packaged in clear bottles for visibility or "what you see is what you believe or experience." Beer is sensitive to sun light or light that produces the light struck or skunky flavor. The cause of light instability was caused by a photolysis of hop bittering compounds and sulfur compounds in beer. To overcome that photochemistry, Miller developed a hexane process to modify the bittering compounds in 1960 to reduce the skunky flavor. Being a chemist, there was the task for us to start with. So, our goal was to make the process environmentally friendly and safe. At the time, the CO_2 extraction and HPLC were just emergent and that made it possible to produce a high quality, environmentally friendly and food safe product. The advent of CO_2 hop extract led us to develop a separation of the hop components: pure alpha-acids, beta-acids and hop oil in an aqueous system. Besides the alpha acids are principal bittering precursors and hop oil are used to flavor the beer, we invented a conversion of the beta acids, a brewing byproduct, into bittering compounds. After the bittering chemistry, we went into the flavor chemistry of hops. The hop chemistry has thus been evolved within the brewing industry. The ultimate outcome of all of this research at Miller thus resulted in them building a hop processing plant in 1987. Most of the key hop chemistry work is performed in aqueous systems. Imagine doing organic chemistry without using any organic solvent.

Beer skunkiness
Take the isohumulone from hops, a B vitamin, a thiol donor then add UV radiation and what do you have? A reminder of what it feels (or rather tastes) like to be sprayed by a skunk. The compound formed is 3-methyl-2-butene-1-thiol (MBT) which, interestingly, is in the same class of some of the smelly compounds (thiols) found in skunk spray.
https://beerandbrewing.com/off-flavor-of-the-week-lightstruck-skunky/
08/10/23

8.3.2 What classes that you took apply to some of the things that you have done in the brewing industry? Why/how?

First, as an undergraduate you took all of the core courses that you could, right? Organic, inorganic, physical chemistry. In graduate school you advanced those three important courses. They are the three basic ones that you have to be very good at it because you have to face them in your work. You need to understand the theories and know how they are applied. I'll give you some simple examples. For inorganic, first you know that when acids neutralize they create a salt. When you deal with organic acids, they can go into aqueous solution by forming salts then you acidify them and you can get them out of solution so that you do chemistry in an aqueous media. In addition, by means of diphasic oil/water system or precipitation, you can get the organic acids out.

8.3.3 What adjustments did you need to make from working in a lab setting to a brewery setting or how are they different?

In a lab you are unlimited by your imagination and operation. You can study almost anything. However, transition from a lab to large-scale work, and you must go through a pilot scale to make many adjustments in order for it to become practical. For instance, acidification is done in a beaker. If stainless steel is used in the plant, HCl is corrosive, H_3PO_4 is unfavored for the environment and H_2SO_4 must be cautiously handled. In an industrial setting, you have to consider all the details, safety, yield, quality and the economics of what you are trying to do. You have to be a little more creative in that regards.

8.3.4 What did you enjoy most about your position at Miller Coors?

My last 10 years before retiring, I enjoyed a freedom and free thinking to do my own research on hop chemistry and hops overviewing and envision of how to apply them toward practical applications.

8.3.5 Do you have any words of wisdom or suggestions for students wishing to apply their chemistry (STEM) degree toward getting into the brewing industry?

Brewing is still an art. There are many opportunities to do science. If you want to work in the brewing industry, a few universities in food science offer brewing courses and researchers with a strong chemistry background that can help you to develop your proficiency in brewing. Always asking "why and how" is a good way to understand and approach things.

8.4 Hops industry

8.4.1 You mentioned that you grow hops. Could you describe the process behind growing hops and the different varieties that you have grown?

Hops are a perennial plant. They grow in the area where there is a lot of sunlight during the growing season. It requires about 12 h of sunlight in the summer time. That is why the best growing region is between the latitudes of 45° and 55°. It is the best zone for growing hops in order to give the best yields. If you grow them at home then you don't really care about the yield as much. They still need plenty of sunlight for them to grow. Growing them is easy. You need a good fertilizer and soil and you need enough water. They need to stay dry above roots and wet below in order to grow very fast. They will grow one feet per day at the height of their growing season. It will take about 3 months. After 3 months they will start flowering and take 1–2 months to finish growing. The flowers become mature and then you need to harvest them right away. You don't want to wait for the flower to open up. If it opens up then it's not protected from the inside and lupulin glands undergo oxidation. At the end of August or early September you start harvesting to pick up all the flowers. That is very labor-intensive work. You take the flowers then and dry them because fresh hop flowers or cones contain around 80% moisture which will ruin the vegetative material during storage. So, you dry them out at 55–65 °C for 8–10 h. After they are dry, you can store them in the freezer until you brew. I have grown some commercially available public varieties, except a specialty Citra hop from my particular relationship with the producer because I discovered Citra.

So you want to grow some hops?
Depending on the source, the optimal growing latitude for hops is going to vary so we
won't debate the optimal growing zones or latitudes. If you're inspired to grow your
own hops a good technical starting point would be USA Hops which services Hop
Growers of America. Your state extension service is another good place to start as they
will have more localized information about growing. It is worth noting that a 2019
report from USA Hops indicates that 96% of the hops grown in North America were
grown in Washington, Oregon and Idaho.
https://www.usahops.org/
08/10/23

8.4.2 What drew you to the hops industry?

I always liked the natural chemistry in plants. Hop flowers or cones are an intriguing
plant material. Most fundamental chemistries of hops have been done in academic
research. Miller was an innovative brewing company and specialized in making the
end light stable hop bittering product. The aspects of hop chemistry made us look toward practical hop research to be implemented in brewing. I started with how to
eliminate the hexane for making this light stable product that drew me to study hops.
This led us to practice hop chemistry in an aqueous system and produced various hop
products to add values of hops that are still followed by the hop industry.

8.4.3 How has the hops industry changed since you first started in it?

For the mainstream beers hops are conventionally added during wort boiling in a kettle for a longer time. It is mainly for achieving the maximum and consistent bitterness. It turns out that the beer does not have much aroma. So, for a lager or a pilsner
that is what it would have. We invented solvent-free hop processing to make advanced hop products for bittering beer and worldwide use. Therefore, bittering hops
or high alpha-acids hops are dominated for all lager or pilsner styles and growing and
breeding high alpha-acids hops are an important variety in the industry.

    During the Miller/Miller Coors yearly hop tour, we often brought back some new
or experimental hop varieties for evaluation. Testing and selection of hops never
stopped because it was the company's strategies for sustainability and consistency of
the existing products and new products. After focusing on the bittering chemistry
(high alpha acid hops), we started to search for aroma hops other than the existing
varieties, but the question was what caused aroma and nobody really knew. Through
our practical hop research, besides hop acids and essential oil, we discovered that the
waste vegetative portion or hop solids also contribute essential flavor, antioxidants
and textures of taste. A goal of my hop hunting was for dual purpose hops that could

maximize the usage of hops and endure Miller/Miller Coors hop strategies. I contacted a hop producer who sent me a few experimental hops for testing. One experimental hop drew my attention, which contributed very unique and exotic aroma and flavor associated with grapefruit, passion and tropical fruits, lychee, mango and citrus notes – these were never perceived in any hop variety to date. The hops are now named Citra. The flavor chemistry of Citra brought worldwide hop researchers' interest and also the interest of craft brewers. On the rise of craft brewing, the Citra hop flavor profile provides an edge for them to break out of the traditional American lager, European pilsner (noble hop flavor) and British IPA. Citra hops took off and became in very highly demand worldwide. The Citra hop flavor has changed the consumers' palates, the hop industry and the brewing industry.

8.4.4 What is the biggest challenge to the hops industry right now?

Labor is always a concern for the hops industry. The industry is attempting to develop more automation for harvesting. The industry does not have an onsite measurement of aroma and flavor but relies on the sensory evaluation of the final beer. Even GC has a difficult time distinguishing those two. The mystery of the black box for hop aroma and flavor in beer is yet to be revealed. The other thing is you do not know what the consumer will want next.

8.4.5 I've heard of some new bittering compounds associated with dry hopping. Any comments or thoughts on that?

First, principal bitterness comes from the iso alpha acids or isomerized alpha-acids. In the past more than 80% comes from those due to aged hops. The bitterness of today's lager or light beer is mainly contributed by iso alpha acids because processing, packaging and storage keep the hop fresh. The potential oxidation side products of humulinones, hulupones and others derived from hop acids are trace or minimal in hops. However, heavy or particularly dry hopped craft brews elevate the concentrations of those side products and so does the bitterness potential. In my "Deep Hops" book, to be completed soon, I have elaborated the mechanism of oxidation of alpha-acids to humulinones."

8.4.6 Can you describe your most interesting contribution to the industry?

Our technologies using an aqueous separation of hop CO_2 extract and making advanced hop products have added flexibility and values of hops for the industry. First in the industry, our semiautomatic aqueous hop processing plant or Watertown Hops

Co. that we built supply not only MillerCoors (now MolsonCoors) but also breweries around the world as well as other applications. A lot of the research comes from those hop products through that. The other one of course is the discovery of Citra hops. It has changed the craft brewery world and people's perception of beer.

Watertown Hops Company

The Watertown Hops Company focuses on the generation of hop products designed to enhance various features of beer including foam, light stability and consistency. The company was founded in 1987 with a primary focus on the light-stable products. Most of the products listed on their website still rely on the CO_2 extraction techniques that were developed in Dr. Ting's time.

https://www.watertownhops.com/
08/10/23

8.5 Professional affiliations

8.5.1 What current or former professional organizations were you affiliated with? Why did you join them?

I used to be a member of ACS. I was once awarded for a chemist achievement in the ACS Milwaukee local section. The ACS is the biggest chemistry organization in the world. I joined this organization because it covers all the chemistry in the world. You can have access to broaden and advance your knowledge in any area of chemistry. I am still keeping up but limit myself to the abstracts of Journal of Agricultural and Food Chemistry, monthly. On the contrary, the ASBC is smaller and focused on brewing — I still have a membership. The ASBC is easier to do networking and exchange ideas during sessions and beer tasting. I have received the ASBC Eric Keen Memorial Award for an excellent scientific publication in hops. I have also received more than 25 US and International Patents in the field that I am in. I really enjoyed my work, particularly in the last 10 years before I retired. I am passionate about my work and like to pass along my experience to the next generation. After retirement, I was hired by the Hop Growers of America to promote American hops and also was one of delegates for ASBC to present hop chemistry in China.

Eric Keen Memorial Award

The Eric Keen Memorial Award sponsored by the ASBC is awarded annually to outstanding JASBC articles. Selection criteria include: technical rigor, originality, number of citations and clarity. Dr. Ting received the award in 2008.

https://www.asbcnet.org/community/Pages/Eric%20Kneen%20Memoria%20lAward.aspx, 08/10/23

8.5.2 What do you enjoy most about going to their events?

I usually went to ACS conferences because there are so many interesting papers, presentations and information outside of brewing to be absorbed in every area. The events were so big and intense at that time I wished that I could enjoy and cover all of them. After I attended ASBC, the events were so focused, sociable, easy networking and relaxed.

8.5.3 How important is it to be associated with other professional organizations?

Yes, so you can exchange ideas or to pick up new ideas and trends. It is also very good for networking. Particularly, if you can publish paper or make presentation, it will help you too.

8.6 Other/miscellaneous

8.6.1 Outside of brewing what other hobbies or interests do you have?

Unfortunately, I do not have a specific one except for growing hops at home. Writing and reviewing papers as well as homebrewing are also keeping me busy.

8.6.2 Is there anything else that you'd like to share about your personal or professional life that you think people might be interested in knowing?

All I would have to talk about is my professional journey. I think you do need to be passionate about whatever you are doing. Learning is endless and knowledge is acceptance, adaption and application.

9 Irena Wise

Winding Path Brewing, Dallastown, PA

BA in Chemistry from Lewis & Clark College – small liberal arts college, primarily undergraduate 2011

Lewis & Clark College
https://www.lclark.edu/
08/29/23

Winding Path Brewing
https://wyndridge.com/
craft-beverages/beer/
08/29/23

9.1 College/degree pursuit

9.1.1 What inspired you to pursue a chemistry degree?

Because I went to a liberal arts college and got a BA, I have more of a background in writing and foreign languages than other chemists might have. Nonetheless, I still have a very strong chemistry background. I was able to do more graduate-level things while in college, such as helping a professor with their research and serving as a teaching assistant in undergraduate labs, because LC didn't have chemistry graduate students.

At the time, I was planning to go to graduate school, so for my undergrad degree I just chose the subject that I enjoyed working on the best. I liked Spanish and English but was also very strong in all of the STEM fields. I liked doing problem sets more than anything else. I liked how chemistry focused on practice problems rather than just reading and writing. I preferred doing my homework as problem sets and in study groups and going to class sessions and doing problems on the board. I just enjoyed working in the lab.

9.1.2 How long did it take you to graduate? Were there any stumbling blocks you'd like to share?

I did it in 4 years. The biggest stumbling block was graduating with honors and doing an honors thesis. There were several times during the thesis – when I was putting in extra hours – where I was wondering, is this worth graduating with honors? Is it worth the extra time I'm putting in? Will it help my career? In the end, I am glad that

https://doi.org/10.1515/9783110798777-009

I did it because of the camaraderie with the other students. There were only three of us working on one at the time, and we all supported one another. I also enjoyed getting to work so independently. I really enjoyed being self-directed. It wasn't an honors college or anything, just an honors thesis that they offered.

National Collegiate Honors Council (NCHC)
Now this truly was a blast from the past. I served as Honors Program director at Angelo State University and once served as a site visitor for NCHC. Collegiate Honors Programs strive to give students and enhanced education that provides them with unique opportunities. Honors courses aim for their content to be more in depth or more in breadth while not being "just more difficult" than their regular section counterparts.
https://www.nchchonors.org/
08/29/23

9.1.3 Did you do any research as an undergraduate? If so, what was the project?

Actually, my thesis tied into a summer research program called the John S. Rogers Science Program. It was for 10 weeks and you received a stipend for it. You could work any 10 weeks, scheduling them around your vacation or your advisor's vacation. I was able to enter that program for two summers and then continue the research in my honors thesis. So, I already had a baseline for writing my thesis proposal when I entered my senior year because I had already been working on it during the summer.

The project itself was very theoretical. It's been so long since I've talked about it that I'm not even sure if it is still the main way that theoretical chemistry research is done. We worked with CASSCF to approximate solutions to the Schrödinger equation and describe orbital functions. We were finding the most likely transition state for various reactions. We strung a bunch of processors together and ran calculations to find the solution to each equation that had the lowest energy. We were lucky because we had the best work environment in the program, with good computers and speakers and chairs. We weren't actually in a lab, standing up and working with chemicals. We were also the most in control of our schedules because things could be running in the background and we could take breaks. We could meet up with the other research students and play soccer during our lunch breaks.

John S. Rogers Science Program
This is a local program offered by Lewis & Clark that is similar to the McNair Scholar Program. Students participate in a 10-week summer research program in the sciences. Participants are trained so that they "have a responsibility to communicate the purpose and results of their work to a general audience."
https://college.lclark.edu/science/opportunities/rogers/
08/29/23

CASSCF
Computational chemistry was never my strong suit. If you're curious though feel free to take a look at this Gaussian site.
https://gaussian.com/cas/
08/29/23

9.1.4 Do you have any words of wisdom for students wishing to pursue a chemistry degree?

The lower-level chemistry classes are so simplified that they tend to not make sense. You just have to take these concepts at face value because of the way they are explained. Just trust that the subject matter is being presented in the way you need to understand it in order to get to the next level. Stick it out for 300 (junior) level because that is where you get the best explanations that provide the best approximation of reality. Those explanations are more satisfying.

9.1.5 How often do you use what you learned as an undergraduate in your current job?

I use it all the time actually. I have an easier time adjusting to new concepts and addressing new problems than brewers without a STEM background. It's probably because I am more comfortable with chemical terminology. For example, there's a concept that is just becoming important to craft brewers that requires knowledge of chemistry. It's bitterness is caused by dry hopping. We're dealing with a newly discovered class of bittering compounds that are actually soluble during dry hopping. We used to think that no bittering compounds were transferred to the beer during dry hopping. Because of my chemistry background, just being able to differentiate between the main bittering compounds, I was able to understand new research in that area. It especially helps when I have to explain why some beers turn out more bitter than expected.

I always tell people that the class that I use the most day to day was a two-credit water chemistry course that I took. It was focused on temperature-dependent solubility of gasses in water. <Flynn: Yeah, like the force carbonation chart.> Just having a really good understanding of the solubility of carbon dioxide in water being pressure and temperature-dependent has helped me the most. My instructor was mostly focused on the rising temperatures in the ocean, but it also applies to beer carbonation.

Forced Carbonation Chart
If you know the temperature and PSI of your CO_2 source then you know the volumes of CO_2 that your beer is going to have using the Forced Carbonation Chart. What the chart does not tell you though is *how long* it will take your beer to get to that level of carbonation. The typical slow method will take a little shy of a week to get to the desired level on the homebrew level.
https://www.kegoutlet.com/keg-carbonation-chart.html
08/29/23

9.2 Movement to brewing field

9.2.1 What was your "entry beer" into enjoying craft beer? Why did you like it?

Definitely Bell's Oberon. I grew up outside of Detroit, and that was the first craft beer that I saw other people in my family drinking. I think it's just sentimental.

Bell's Oberon
Oberon is a wheat beer with a 5.8% ABV that has a simple four-component ingredient list. It is described as "Citrusy, easy drinking and refreshing, Oberon Ale is sunshine in a glass."
https://bellsbeer.com/project/oberon-ale/
08/29/23

9.2.2 Did you get started with homebrewing?

I did homebrew a little bit. I homebrewed during my senior year in college. I had no classes on Thursdays that year, so sometimes I would work on my thesis, sometimes I would go snowboarding and sometimes I would homebrew. Our homebrew setup was so basic – just a big pot on the stove. It took a lot of time for the wort to cool down from boiling to yeast pitching temperature, so I would just do my homework then.

9.2.3 What was your first recipe?

A red ale.

9.2.4 Did you ever enter a recipe into a brewing contest?

Not as a homebrewer, but I have entered plenty as a professional one. We've entered GABF every year that I've been at Winding Path.

Red ales
If you take a look at the 2021 BJCP guidelines you'll see that the two specific Red Ale
subcategories are the Irish Red Ale and the Flander's Red Ale. The only thing these two
really have in common is that they are both red in color and have a higher malt profile.
It is interesting that despite there only being two subcategories, Red Ale is referenced
23 times in the guidelines.
https://www.bjcp.org/bjcp-style-guidelines/
08/29/23

9.2.5 What was your biggest disaster as a homebrewer?

No exploding bottles or anything. All the beer turned out pretty good. I realized I
could spend 8 h making 5 gallons of beer or I could get paid to spend 8 h making 1,000
gallons of beer. So, I switched to professional brewing pretty quickly.

9.2.6 Were you part of a homebrew club? Which one? How long a member?

No. But I was living in Portland, Oregon.

9.2.7 What was your biggest batch as a homebrewer?

Just five gallons.

9.2.8 Did you ever brew a beer for an event (wedding)?

Yes, for my own college graduation.

9.2.9 Prior to getting into brewing what else did you do after graduating?

Nothing. I worked for free at a brewery in Portland during my senior year. I worked
with a brewery, Lompoc Brewing, which no longer exists. They bottled once every 2
weeks. On those days volunteers could come in and help and get paid in lunch and low
fills. Lompoc only bottled twice a month because they used a contract bottling company
and didn't have their own bottling line. This contract bottler would come in and set up
their bottling line and hook up Lompoc's tanks to their filler. Volunteers would stand at
the end of the line and take full bottles and put them in cases. So, I had been doing that.
After I graduated they had a job opening in August and that was how I got started.

9.2.10 Would you recommend brewing to other chemists out there and why?

Yeah I would, especially because other pathways open up once you learn brewing and cellaring. There are several new areas that are growing, especially on the quality side, where it's important to make sure that everything is sanitary. A lot of brewers are taking their knowledge of sanitation, packaging and microbial stability and moving into NA beverages or "wellness" drinks. There is a lot of work available for brewers to brew and pasteurize all these new beverages. It's not just beer production that has jobs for chemists, but it is other areas as well. Let's not forget that hop producers and yeast producers need their own chemists, too. For example, Imperial Yeast just opened up a facility in Philadelphia.

9.3 Current position

9.3.1 What are your current job responsibilities and tasks?

I am currently the head brewer and quality manager. The farm where I work started out as a cidery, but it now produces both a line of ciders and a line of beers. I work closely with the cider master – he handles all the cider design and production. I handle beer production in general, yeast propagation, raw materials sourcing, writing brand descriptions, and explaining the brands to our sales team, among many other things. I also handle contract brewing and working with clients, adapting their recipes to our equipment.

9.3.2 What classes that you took apply to some of the things that you have done in the brewing industry? Why/how?

The water chemistry course I mentioned before applies to the brewing industry. That class covered gas solubility as it relates to temperature and pressure.

Water chemistry
There are a few options out there for water brewing calculators and many of them focus on the brewer declaring what style of beer they are brewing (e.g., Hoppy Amber) and inputting their known ion values for the source water. My personal preference is to use RO water and build my water profile from there. It has produced the most consistent results for me at least.
https://www.brewersfriend.com/water-chemistry/
08/29/23

9.3.3 What adjustments did you need to make from working in a lab setting to a brewery setting?

Being on your feet and working long days and building up your stamina. Breweries get hot on the brew deck and cold in the packaging hall, and you might be on your feet for 10 or 12 h if something goes wrong that day. A lab setting is temperature-controlled. You can take breaks. But in brewing, once you start a brew, once the water hits the grain, you have to finish it out. If the lauter gets stuck or if the boiler shuts down, you just have to work through it and accept that your day will be longer. You're just stuck there. You may have to eat lunch at the boil kettle in between hop additions. I think it's worth it. I find it satisfying to use my brain and my body at the same time. But that is the biggest adjustment – being on your feet all day.

9.3.4 What do you enjoy most about your current position?

I enjoy the variety of responsibilities and the freedom. It also gives me the chance to learn new things. I have plenty of experience brewing and handling the technical side of beer production. But, this is my first time sourcing raw materials or contract brewing. So, I'm learning a lot about supply chain and how to make the same beer on two different systems.

9.3.5 Do you have any words of wisdom for students wishing to apply their chemistry (STEM) degree toward getting into the brewing industry?

Definitely study water chemistry. Learn your physics. Brewing has a lot to do with fluid dynamics because a lot of our job is working with pumps and pipeworks. For example, you need to understand the relationship between cleaning and laminar flow. So, don't skip out on your entry level physics. In addition to studying, do informational interviews and ask brewers about their jobs. They'd be happy to sit down and have a beer with you. We consider it a perk of the job.

9.4 Experiences in the brewing industry

9.4.1 How would you describe your leadership style?

I took a class on this in 2019. It was a women in leadership class through Cornell online. I received a scholarship for it supported by the Pink Boots Society and Yuengling.

My leadership style is definitely teaching. I'm always interested in the personal and professional development of my coworkers. I try to give them a lot of indepen-

---

Cornell Leadership Certificate
There are several options for getting training in leadership including the Cornell certificate program. I'm by no means an expert in the craft, but what I will say is consistency, communication and respect go a very long way in any industry.
https://ecornell.cornell.edu/certificates/leadership-and-strategic-management/leadership-essentials-certificate/
08/29/23

---

dence and then review what they have done. We go over why each thing was done a certain way and whether they're going to keep doing it that way.

9.4.2 What motivated you to become a master brewer?

I got the certification because I thought that I had plateaued in my career at the time. I was a very skilled operator but I wasn't moving into the management level, so I wanted a formal brewing education to give me an edge. I was continually being assigned shift work, having to work third shift, and I hoped that the certification would help me move into a management position.

9.4.3 What are some unique strengths that you think women bring to the brewing industry?

You'll hear this word tossed around quite a bit – balance. I think women bring balance to the brewing industry. I think women are willing to step back and say "I'm getting burned out. I need to protect myself and step away from this particular problem." That protects them and their staff in the long run rather than causing burn out and causing people to quit or switch breweries every 2–3 years. It's being able to stop and say, "This project isn't a good use of our resources right now."

9.4.4 Have you had any negative experiences in the brewing industry that you would like to share?

Nothing in particular. My most negative ones have been routinely being ignored when I am with male brewers. In 2017, I went to a conference with a male brewer. We had the exact same position; we were counterparts. We were walking through the trade show. I didn't really notice anything unusual. When we left, though, my colleague commented on how awful it was that everyone greeted him and ignored me. I didn't even notice and just thought that they were unfriendly. That is one thing that I didn't understand when it would happen to me at events such as beer festivals. People would say something to my male counterparts and not me. They would not introduce themselves to me and I would think "Did you think I was here as someone's girlfriend? Even if that were the case, why would you not tell me your name anyway?"

9.4.5 What are the various ways that you keep yourself grounded and you use to take care of yourself?

Like I mentioned earlier, being willing to step back and say that a certain task does not need to be done today, or that some task will not give us enough of a benefit for the time and effort put into it. Besides that, I use mindful breaks throughout the day. Before I go into the brewery, I stop outside my car and stand still for 20 s. It's something that I learned from healthy eye tips. If you look at a screen for 20 min, then you need to look away for 20 s. I moved that into the brewery and other parts of my life. If I've been staring at the brew house HMI for a long time, I'll take a look around the rest of the brewery to get a break. In the brewery, as long as you're not looking at your phone, people know that you are listening to the equipment and therefore keeping track of the environment. So, I try to take breaks from staring at the control screen by checking on the rest of the cellar floor.

Human-machine interface (HMI)
HMI sounds real fancy, right? These helpful tools afford brewers very traceable, precise data and interaction with their instrumentation. Processes can now be monitored and adjusted from a distance.
https://www.portlandkettleworks.com/hmi-control-panels/
08/29/23

9.4.6 Do you have any advice that you would give to your younger self before getting into the brewing industry?

Be ready for the physical aspects of the job. Be ready to use pallet jacks and be on your feet for a long time.

9.4.7 Do you have any tips or tricks for maintaining a positive work-life balance?

Yeah. My best trick is to count events as part of your day. One of the hardest things about being a brewer is that we are expected to brew for 8 or 9 h and then attend tastings. The best way to handle that is to start your day later when you have an event in the evening.

9.4.8 Why do you think there is such a relatively small number of women head brewers in the industry?

I'm trying to think of all the women that I have met that are brewers and those who are not but are surprised to see what I do. I know most people are surprised to see what brewing entails. Most women just do not understand the day-to-day schedule of a brewer, so they don't pursue it. I think it also comes from the unfortunate perception of beer in our society as just being domestic light beer. Why would a woman want to go through the trouble of lifting hoses and cleaning kegs if all she is going to do is make a more expensive version of domestic beer? Diversifying the industry starts with teaching people what beer can be. If we can show people that beer can be a full-flavored nutritious beverage, like it was 200 years ago, then maybe people would be more interested in the job.

9.4.9 Describe your toughest day in the brewing industry so far

There have been quite a few, usually when something broke or went down. The one that comes to mind is when the glycol lines went down. The glycol pipework just cracked at the outlet of the pump, resulting in all our glycol being pumped onto the floor. I was in the office training somebody, and we heard this huge bang and splash. We went into the cellar and the floor was under an inch of this slippery green liquid. For a moment we didn't even know who to call because we did not even know what the liquid was. We knew it was not beer, and it did not look or feel like CIP chemical. We didn't know if we were standing in something that could hurt us until we figured out it was just glycol. That's the worst, when the cooling system goes down. Glycol is an expensive compound to have gone down the drain. Then you have to explain to your boss how you lost $2,000 worth of refrigerant, and you're about to ruin all of your fermentations if you don't get the system back online. I think the other worst days have been during dry hopping, when the dry hoppers have exploded on me. If you are using a special dosing chamber, it can clog and back up when you try to empty it. Then, you don't realize how full it is and end up spraying beer out the top and getting it all over the cellar. We've also had the hops cause CO_2 nucleation, which causes the beer in the tank to turn into a geyser, and again you get covered in beer. That's another of the worst days. You're covered in beer, you lost most of your dry

---

Glycol chillers – G&D Chillers
The manufacturer website boasts that they serve over 2,400 breweries. This is actually a very helpful site that provides installation videos, FAQs and product manuals. Definitely worth a look if you're planning to get a glycol chiller or want to learn more about them.
https://gdchillers.com/applications/breweries/
08/29/23

---

hops to the floor, and you don't know how many hops to put back in. Again, that's an expensive loss. Those hops cost $16 per pound!

9.4.10 What are the major differences between QC/QA in a cidery and QC/QA in a brewery?

On the cider side, quality staff deal more with preservatives because cider often has residual fermentable sugars in it. I was not familiar with the technique of back-sweetening before working at a cider operation, but it's quite common and means that cider needs to be stabilized with pasteurization or chemical preservatives. Back-sweetening is when the cider maker ferments the cider to dryness and then adds some unfermented juice or other sugar back to it. That allows the cider maker to adjust the final balance and sugar content. Since there is usually 1.5–2.0 Plato being added back, you have to be very careful to use chemical stabilizers to stop more fermentation. So, we are always testing sulfite levels to make sure that we are in the exact range where sulfur will be undetectable and yet suppress microbial activity. It's quite a bit more pH adjustment as well because those stabilizers only work at the lower pH of ciders and wines. We use malic and citric acid to adjust pH in the cidery. On the other hand, you don't have to worry as much about staling and oxidation with cider. I'd say that's the big difference. Cider quality is all about pH control and preventing refermentation; beer quality is all about oxygen exclusion and preventing staling.

9.5 Professional affiliations

9.5.1 What current professional organizations are you affiliated with? Why did you join them?

MBAA and ASBC. I joined mostly for the online resources. They also have forums where you can talk to other brewers. Last year they had a golf outing where brewers from all around the region came and spent the day on the links together.

9.5.2 What do you enjoy most about going to their events?

Hearing what else is going on with other breweries. Are they having similar issues? What kind of beers are they making? I enjoy learning from each other and seeing how others are handling issues that pop up across the industry.

9.5.3 How important is it to be associated with other professional organizations?

It's very important to meet with others and learn from other people out there. Some of them have mentorships or internships and job opportunity postings.

9.6 Other/miscellaneous

9.6.1 Outside of brewing what other hobbies or interests do you have?

Right now it's just childcare. Also I like exploring the surrounding area and hiking, hanging out with my dogs.

9.6.2 Is there anything else that you'd like to share about your personal or professional life that you think people might be interested in knowing?

I do have a master brewer's certification. I spent six months in Germany in 2018 to get it. That's the closest thing I have to a graduate degree. I studied and took my exams at the VLB in Berlin. I highly recommend them. The continuing education available to brewers, like the master brewers' course at the VLB, is varied and accessible even after one has started a career.

VLB Berlin
Versuchs- und Lehranstalt für Brauerei has been in operation since 1883. It is affiliated with the Berlin Institute of Technology. In addition to providing consultations and services they offer various training and educational opportunities including: microbiology, sensory, brewmaster certifications and in-house training.
https://www.vlb-berlin.org/en/home
08/29/23

---

10 Martin Beaulieu

BS, Biochemistry, Université Laval, Quebec, Canada 1993

PhD, Physiology- Endocrinology, Université Laval, Quebec, Canada 1998

Quantum Brewing, San Diego, CA, Owner/Brewer

Université Laval
https://www.ulaval.ca/
08/24/23

Quantum Brewing
https://www.quantumbrewingsd.com/
08/24/23

10.1 College/degree pursuit

10.1.1 What inspired you to pursue a chemistry (STEM) degree?

I love biology, medicine in general and I hated math, so I said that there are no jobs being a biology teacher and I went with biochemistry. The thing is that the first class that I had to take was a math class. I told myself that I better be good at math because it is everywhere.

10.1.2 How long did it take you to graduate? Were there any stumbling blocks you'd like to share?

It took me 3 years for my undergraduate and 5 years for my PhD. We'll talk about my research in a bit, but my PI found these metabolites in the blood and wanted to clone the genes associated with making them, right? Two other students tried and failed before me and another scientist tried and failed and then I was asked to try to do the same thing. So several people, including those three, were asking me what I was going to do if I fail as well. People did not want me to succeed but I am a winner.

10.1.3 Can you tell me about the research you did while in school?

I used biochemistry and molecular biology and cell biology techniques to research and understand the metabolism of steroid hormones in cancer. Many cancers are de-

https://doi.org/10.1515/9783110798777-010

pendent on hormones which is why some doctors advocate castrating men when they have prostate cancer. I was very successful with my PhD. My PI was actually an inorganic chemist and in his work he found that many of the metabolites of steroid hormones in blood were glucuronide metabolites. There are two things that happen to steroid hormones. First, they are made in the adrenals and the sex organs and then there is a series of modifications to the cholesterol molecule. The molecule is modified by the P450 enzymes to make testosterone, for example, and then there is another P450 located in the skin that transforms it into dehydrotestosterone which is the active metabolite. After that the concentration is fine-tuned because you have to have the excretion component as well. I discovered and cloned the genes that are associated with the excretion of many of those hormones – the genes were for UDP – glucose transferase enzymes. They were well known by the pharma industry because they metabolize drugs. My PI was finding steroid hormones with the glucuronide function in the blood so he wanted to find the enzyme that did it and I actually found it. The steroid hormone is lipid-soluble and does not want to go in the urine or bile. Once you add the glucuronide on it you see a big change in solubility for both the urine and bile. Then the kidneys can excrete it. Jaundice is another example of this application. When you have a deficiency in the bilirubin transferase enzyme associated with taking care of bilirubin you end up with jaundice. Bilirubin is a side product of red blood cell breakdown. You know when babies turn yellow sometimes you have to put them under UV light for a while so that they can break the bilirubin down. The liver needs to develop in those babies so that it can take care of the bilirubin. It's the same enzyme system that takes care of steroid hormones. I used sequencing, cloning and biochemistry techniques to characterize the enzymes. Sometimes I would make stable cell line clones of the enzymes where I would put plasmids with the genes to study their function in cells. We published over 17 articles on that.

Newborn jaundice
Bilirubin is a byproduct of red blood cell replacement. A normally functioning liver will break this down for the body. The placenta passes bilirubin off to the mother so that it can be metabolized, but newborns sometimes have issues with processing it. In this case, bilirubin builds up in the newborn giving their skin a yellow tint.
https://medlineplus.gov/ency/article/001559.htm
08/24/23

10.1.4 Do you have any words of wisdom for students wishing to pursue a chemistry (STEM) degree?

You need to have a strong background in math as well as bioinformatics for biochemistry. Everything associated with that and actually some other fields is being able to use big data.

10.1.5 How often do you use what you learned as an undergraduate in your current job?

Brewing is a lot of biochemistry. It has barley and wheat and malt and is full of polysaccharides. We all know that polysaccharides are just a whole lot of glucose molecules joined together. It is more stable to have glucose in that form for seeds to grow. In the spring the seed has that amylase become active and it helps to break down those sugar chains so that the seed can grow. It cleaves those molecules into glucose so that it can grow until it reaches the sun. Now with brewing we trick the seed into thinking that it is spring through malting to get it to start making the enzyme then we dry it out and it is ready for brewing. I like to tell people that brewing is a biochemistry lab within a microbiology lab setting. You do a major biochemical process and then you ferment it. I really do love doing it and every time that I brew I can visualize what is going on at the molecular level. You can modulate the amount of fermentable to unfermentable sugars as well using different mash temperatures and times. That is very important and is actually Biochemistry 101.

Malting process
This is a great read if you want to understand the malting process. There are several steps associated with malting: steeping (soaking grains in water), germination (allow the start of grain carbohydrate conversion), drying with kilning or roasting completing the process.
https://www.brewingwithbriess.com/malting-101/malting-process/
08/24/23

10.2 Movement to brewing field

10.2.1 What was your "entry beer" into enjoying craft beer? Why did you like it?

When I was doing my PhD we went to a conference in Los Angeles in 1996 and we went to a place where they had a beach. My PI told me there was a nice hotel to go to there so we did that before we came back. Across the street there was a Ralph's and we went in there and saw this beer and it was an IPA. It was Sierra Nevada Pale Ale and we got a case of it and really enjoyed it. We then got another case of it later. The beer had so much flavor.

10.2.2 Did you get started with homebrewing?

I did actually. I brewed in college and it is an interesting story. When I was a second-year undergraduate one of my professors told us to develop an enzymatic assay with a $100 budget and we could analyze whatever we wanted to analyze. We went ahead and made beer using a beer kit. We bought the alcohol dehydrogenase enzyme as well. We made a batch of beer, took samples throughout fermentation, froze them and then analyzed them for alcohol. We made a nice chart showing the increase in alcohol over time – that was the science part. We got an A for the project. My classmate was not interested in brewing so I kept the kit and it was cool because I actually got the kit for free and, even better, it was paid for by the university. Then I kept making beer. At the time it was just to make cheap beer for parties. I'd make a 5 gallon batch. It was like $10 for the extract and a pack of yeast and I would brew beer. The thing is that it would only last one evening. I wanted to make it last longer so I just added more sugar since sugar was cheap. I did not care too much about the taste. The sugar would boost the alcohol, but it didn't help how long the beer would last – still just one evening. It was unbalanced and way too much alcohol for the base beer and was not a good thing. I kept making beer at home. One day when I was in Ottawa with Abbott I realized that hops grew well there due to the soil and climate. I started to grow my own hops since I loved both gardening and brewing. I had Cascade, Mt Hood, Magnum for example. You know why Cascade is so cheap? Because the yield is crazy compared to other hop varieties. It is also disease-resistant. It made more hops than the other five varieties combined. In the fall I would usually make an IPA and I would take the hops straight from the vines and taste it during the boil to figure out how much to add.

Cascade hops
These hops provide a great grapefruit flavor/aroma profile due to the myrcene oil that is present in them. They provide a "balanced bitterness" to beers but are often used in the aroma and flavor additions. When I was making an Irish Red once the hops I ordered did not come in the delivery and I did not realize it until it was too late. Not wanting to have a grapefruit red I used some Cascade hops in the bittering addition and the beer turned out fine due to their characteristic bitterness.
https://yakimavalleyhops.com/products/cascade-hop-pellets
08/24/23

10.2.3 What was your first recipe?

So, in Canada, Labatt Blue was my favorite beer. I got a kit that was a clone of it and that was my first beer. My friend's last name was Lapointe and my last name is Beaulieu so we called it LaBeau Dry.

Labatt blue
This is described as the "best selling Canadian beer." Hallertau hops and two row malts are main ingredients. It is described as having "a clean refreshing taste with distinct hop aroma, delicate fruit flavor and a slightly sweet aftertaste."
https://www.labattusa.com/product/labatt-blue/
08/24/23

10.2.4 Did you ever enter a recipe into a brewing contest?

No. I don't really care about my beer reviews. I make beer and people love it and that is good for me. I hear people say that they do not like reds or ambers and I tell them that they have never tasted mine. I give them a sample and they ask for a pint of it.

10.2.5 What was your biggest disaster?

It is a learning experience for me. In the winter I was home brewing on the stove and I got a boilover. I cleaned it and some of it went between the two windows and my wife was so mad. I ended up getting kicked outside for brewing because of that. It's kind of funny though because in the winter in Canada it is very cold with 3 or 4 ft of snow. I would just take my kettle and put it in the snow to cool it down and it was cool in 10 min.

10.2.6 Prior to getting into brewing what else did you do after graduating?

In 1998, I did postdoctoral research in Los Angeles and got a grant from the Canadian government to do my work. I did my postdoc at the City of Hope Medical Center where I worked on mutation detection. The biotech boom started in the late 1990s and since I loved California I got recruited to a company called Sequenom in 2001. I worked there for 5 years developing genotyping technology studying polymorphisms and using PCR. I was the lead of that and the technology is still used today. The company was sold to Labcorp. Then I moved into molecular diagnostics and worked for Hologic for a short time and was then recruited to DiagnoCure back in Qubec to be director of R&D for molecular diagnostic development. There we created staging tests for colon cancer. The issue with colon cancer then was that when the surgeon would resect part of the colon they would have the pathologist look at the margins and the lymph nodes from the mesentery to determine whether they got all of the tumor. They also want to determine whether it is stage 2 or stage 3 or if it metastasized to see whether they need to use chemotherapy if they find it in the lymph nodes. Approximately half of the time they miss the diagnosis. It is not that they do not know what they are doing. It is just that there could be up to 50 slides that have to be reviewed and even then they are still relying on some degree of sampling to correctly identify it. We worked on a PCR test for a gene that is only expressed in the colon. So you could test lymph nodes and if that gene is there then you know you need to do chemotherapy. The challenge was to use an RNA extraction method from the paraffin fixed tissue which is really hard because the process breaks the RNA up – it is very challenging. We found out that if you make a very short amplicon you can still detect the majority that was there initially. We acquired the rights to the gene and started a 3-year program to develop the extraction method and the test. In 2006 we started doing that and in September 2008 we were going to the stock market for funding and, of course, the stock market crashed. Timing is everything so I could have been retired way back when. I asked the CEO whether we were going to go to market and was told that it was not a good timing. Around that time I got a call from a recruiter at Abbott labs and it was not that I was unhappy where I was, I just figured that I might as well go and do the interview. In October 2008 they told me that they could start me in January. I went back home and figured that I needed to tell the company I was working for. I was going to give them a month's notice – start of December. A couple of weeks after I got the offer they came to me and said that they were letting the team and me go. Luckily, I already had the other job offer. I did not tell Abbott right away and a little later I told my new boss about that and he said, "How did you manage to get that lucky. I tried that many times." I would have rather finished through with that test because it was very good. Anyhow, I worked at Abbott for 5 years in Ottawa and they had the point-of-care division that I worked in as an R&D manager. We developed cartridge-based assays. These are designed so that the nurse can do the test right in the patient room. For example, when people went to the E.R. with a suspected heart attack they would have to draw blood and send it to a lab for analysis to verify. That would take over an hour to get the results. The cartridge test would take 10 min. So it would save not only time but also money. I worked there almost 5 years and my wife wanted to go back to California. There was a company called Regulus Therapeutics that looked at using micro-RNA for biomarkers. It is very stable in the blood and it can be used as a profile for detecting disease. I worked there for almost 2 years and things did not work out too well. I was tired of the corporate and Biopharma and the uncertainty. I had been brewing beer at home since College and decided to go that route.

Cancer staging

Many of us know of at least one person who was diagnosed with cancer. They will sometimes share what stage they are in. There is the TNM system, the five-stage numbered system and the category system that is used by cancer registries.
https://www.cancer.gov/about-cancer/diagnosis-staging/staging
08/24/23

Cartridge assays – A1C cartridge kit
When you have a physical examination your physician will examine your blood glucose from your blood labs. If it is high, they may have their nurse do an A1C test in the patient room to measure your glycated hemoglobin. Fortunately, there are cartridge kits like the referenced one that can be done during the same visit. Glycated hemoglobin is a longer term measurement (e.g., 1 month) of the amount of glucose in your blood.
https://www.globalpointofcare.abbott/us/en/product-details/afinion-hba1c.html
08/24/23

10.3 Current position

10.3.1 What is your favorite and least favorite process to do and why?

My favorite part is getting the mashing started. It is exciting and it smells really good and it is good to know that the biochemical processes are occurring. My least favorite is the cleaning. There is so much cleaning associated with brewing. When people ask me if they can come to brew here the first thing I do is put them on washing kegs. I tell them it is 90% cleaning and if they graduate from that then I have them do kettles and then I let them brew with me and I might let them brew a recipe for me.

Some of the brewing certificates that are out there are not good because all they focus on is theory. You know when I ask those certificate holders what beers they have made and they tell me they have not brewed a beer I'm like "what the heck?" You invested so much time and money in this certificate and they never had you brew a beer or you never got a kit to make a beer? What good is that if you do not have the hands-on experience? There are some good ones out there that do give them the experience they need though.

10.3.2 What adjustments did you need to make from working in your prior lab settings to a brewery setting?

Actually at the end of my career it was not much lab work and simply just politics. I was missing working in a lab at that time actually. Many things that I did in my career were biochemistry too so I was just excited to make something that was robust and tasted good. One great advantage for a scientist who does brewing is that you know to use lab books and take great notes. My PI convinced me to write everything in my lab book. To convince me why I needed to do that he used a story of how a particular drug was discovered – the guy actually discovered it by mistake. He was doing organic chemistry to make it and he did a batch and came up with something much quicker. He could not reproduce it. He looked at the other side of the page where he was doing calculations and learned that he had made a mistake. I write everything down because of that. I told my people to not cut and paste things into their lab book. It took a while for those to learn how to do that. As another example, I made a batch of my stout that was just incredible. It was different and way better. I forgot that I had to substitute one of my roasted malts for a UK version and it was so lucky. It converted the stout from a good beer to a great beer. Other brewers probably would not have picked that up.

Notes maketh the scientist

Martin gives some excellent advice to scientists about using a lab notebook to keep notes. These allow you to document things immediately, to track data and to refer back to when there is a question regarding methods or materials that were used previously. Critical components to a lab notebook entry are date, protocol being used, number of samples and results. On the academic side, notebooks should rarely leave the lab except when manuscripts or other professional communications are being generated.

10.3.3 What do you enjoy most about being the head brewer there?

I can do what I want. One of my best moments with brewing is when I realize I don't have to go to my boss or a board for permission to brew a beer that I want to do. They might have even killed the idea. Here I get to choose a beer and make it tomorrow. For example, I was eating dinner with my kids at a burger joint and I decided that I wanted to make a cream ale and I wanted to make one with blueberry. I told my kids and they were 5 and 9 and they looked at me funny, but I felt so good about doing it. I call it blueberry pi with the π sign. I love being able to do what I want to do with beers.

How many digits are there in pi?

Your math teachers likely told you to use 3.14 in calculations involving π, but the fact is there are much more digits after that 4. In 2022, scientists in Switzerland calculated over 62 billion digits in π and this number will likely go up as computing capabilities increase. $-1^{0.5} 2^3 \sum \pi \ldots$ and I liked it!

https://www.guinnessworldrecords.com/news/2022/3/new-value-of-pi-calculated-by-swiss-university-at-over-62-billion-digits-694748

08/24/23

10.3.4 Do you have any words of wisdom for students wishing to apply their chemistry (STEM) degree toward getting into the brewing industry?

You don't need a chemistry degree to be a brewer. First though find a beer kit and make a beer on the weekend. It is cheap and you can learn from doing that. Go and volunteer at a brewery and tell them that you are studying chemistry and you'll do what they want. I would really like to have someone with a science degree because I know how they think. Do not waste your money on a brewing certificate right off the bat. Do things at a brewery or with a kit because you will learn so much more for so much less investment.

10.4 Experiences in the brewing industry

10.4.1 How would you describe your leadership style?

I would say that it is trust but check. I understand that different people have different learning styles. I do have protocols on paper and I let them read it, I do it in front of them and let them do it and then they are good to go. It is important because it is my business. I am a teacher so I trust but check.

10.4.2 What are the various ways that you keep yourself grounded and you use to take care of yourself?

You'll laugh, but I drink a beer when I feel anxious. Nobody is going to fire me for that. I bartend a lot and I know people from every background imaginable. I tell my wife she can't divorce me because I know three customers that are family lawyers and would do it for beer. I have a mechanic that will come to my brewery and change my brake pads out for beer. That is the nice social part of brewing and owning a brewery.

10.4.3 Do you have any advice that you would give to your younger self before getting into the brewing industry?

If you make a bad beer get rid of it. You can have 12 beers on tap and 11 of them can be great. That one bad beer will cost you so much more than the cost of making the beer. Good news travels, but bad news or information travels so much quicker. It is hard to do convince yourself to do that though.

10.4.4 Anything else you'd like to share about brewing in San Diego?

It is very competitive. The number of craft beer drinkers has dropped quite a bit. People were locked away due to Covid and worked from home and it was the worst thing for small business. I lost a lot of my after-work business. Also, we are now competing for a smaller pool of customers.

Yeah, when I first left San Diego there was like seven or eight breweries. When I came back 8 years later there was like over 80 beers and it really surprised me. I discovered IPAs when I came back and thought that I would miss the boat. So with me being unhappy with my new job I bought this little brewery and I loved the name. They had some quality issues so I got it in 2016 and decided to take a chance and make better beer. It took about 6 months to get people back. The first 3 years kept getting better and better so in 2019 I bought a new fermenter. I was pretty much maxed out in brewing capacity. I was approached by a group to open up a tasting room downtown. They were going to give me money to help me do it. We had a lease agreement and everything was ready to go in February 2020. Now at that time we were denying that anything was wrong (Covid). I was brewing and it was slow and I turned on *CNN* and was reading my lease agreement. Something was telling me to be careful. I saw them washing the streets in Italy and they were dressed up and disinfecting things. I realized that it could also happen here and I would be in real trouble. I told the guys to pause it for a minute. That was the most important decision that I ever made for the brewery.

10.5 Professional affiliations

10.5.1 What current professional organizations are you affiliated with? Why did you join them?

The San Diego Brewer's Guild for sure. The thing is that I do not have time to attend the meetings. They were very useful during the pandemic and communicated what was going on and when the government was making decisions. They lobbied for us. I am not a member of some of the other ones.

10.5.2 How important is it to be associated with other professional organizations?

There is a good networking opportunity. I know a friend who is a chemist who met somebody that he is now working with. I actually know another one working at a brewery.

10.6 Other/miscellaneous

10.6.1 Outside of brewing what other hobbies or interests do you have?

I got into smoking meats during Covid. It was a similar challenge. I love cooking but it is very exciting because there is so much chemistry in smoking. Now I do it twice a month. I have a great smoker and a pellet grill that I use. My wife told me I could not get another grill.

10.6.2 Is there anything else that you'd like to share about your personal or professional life that you think people might be interested in knowing?

I am a pilot. I was in the Air Cadets in Canada and I was good at school and took some tests and I was able to become a glider pilot. Then I became a Cessna pilot. People asked me if I was going to be a bush pilot. I looked it up and realized that on average each of those has at least two crashes per career. That is not good and I was not good with navigation either. The instructor would ask me to identify one of three lakes on a map and I could not do it. There was no GPS at the time of course.

Canada Air Cadet League
More information about the Royal Canadian Air Cadets can be found at the link. Air Cadets are trained in areas such as "drill, survival, principals of flight and navigation."
http://www.aircadets.ca/
08/24/23

Section II: **The biochemists**

11 Grant Chandler

Hendrix College, Conway, AR, Biochemistry and Molecular Biology, 2012

Lost 40 Brewing, Little Rock, AR, Brewery Operations and Personnel Director

Hendrix College
https://www.hendrix.edu/
08/07/23

Lost 40 Brewing
https://www.lost40brewing.com/
08/07/23

11.1 College/degree pursuit

11.1.1 What inspired you to pursue a chemistry (STEM) degree?

I had a vague notion of becoming a doctor or otherwise working in the health care industry. I was never super serious about that particular plan. Actually, plan is not the right word, maybe excuse is a more accurate description. I think Biochemistry was simply the subject I was most interested in at the time, and health care was how I justified studying it.

11.1.2 How long did it take you to graduate? Were there any stumbling blocks you'd like to share?

It took me just over 4 years. Halfway through college I realized that health care was not an attractive industry to me. I did not experience any particular stumbling blocks for graduating, but I did graduate without much sense of direction or how to best use my education.

11.1.3 Did you do any research as an undergraduate? If so, what was the project?

Yes. I researched the enzymes and pathways involved in alcohol metabolism and how healthy liver cells are damaged by the consequent cascade of reactive oxygen species when metabolizing ethanol.

https://doi.org/10.1515/9783110798777-011

Reactive oxygen species and alcohol
Reactive oxygen species have been thought to be an underlying reason behind organ damage associated with alcohol consumption. When ethanol is consumed it is metabolized by alcohol dehydrogenase and the cytochrome P450 system. Reactive oxygen species result from this metabolism and generate oxidative stress in cells.
https://doi.org/10.3390/antiox12071425
08/07/23

11.1.4 Do you have any words of wisdom for students wishing to pursue a chemistry (STEM) degree?

If you don't know exactly what to do with a STEM degree I would err on the side of engineering or an applied field with the highest earning power. It appears more difficult to move from research to engineering than vice versa.

11.1.5 How often do you use what you learned as an undergraduate in your current job?

Very often. It does not always occur to me that I am using what I learned as undergrad, but I do use it on a daily basis.

11.2 Movement to brewing field

11.2.1 What was your "entry beer" into enjoying craft beer? Why did you like it?

In college I visited the historic Guinness brewery and drinking Guinness in Dublin was a very joyful and a delicious experience. It made me rethink beer. That memory was very specific, but drinking culture in Europe is much more festive than in America. In the UK, pubs are often owned by specific breweries, and I much enjoyed exploring different pubs and the distinct beers they served. That experience at Guinness really topped it off though.

Guinness
Guinness is one of the most, if not the most, famous stouts in the world with an ABV of 4.2%. What many refer to as the "Irish Pint" was first brewed in 1959 in celebration of the 200th anniversary of the brewery. It is noted to have "hints of roasted coffee aroma" and is "smoothly balanced with bitter and sweet notes." While everyone believes it is black in color, the true color is dark ruby red.
https://www.guinness.com/en-us
08/07/23

11.2.2 Did you get started with homebrewing? What was your first recipe?

Yeah. It was a Moose Drool clone that is a brown ale by North Sky brewing. It was an extract recipe.

11.2.3 Did you ever enter a recipe into a brewing contest?

Yeah, and I received quite a few awards too. Consistently medaling was the largest suggestion that I might could turn brewing into a career.

11.2.4 What was your biggest disaster?

I probably worked too hard at it, and my health suffered for a time. I don't know if I have an answer other than that. I was definitely not doing other things because I was brewing all the time.

11.2.5 Were you part of a homebrew club? Which one? How long a member?

I participated in the Central Arkansas Fermenters club for a few years.

Central Arkansas Fermenters
Based in Little Rock Arkansas, this homebrew club boasts to be one of the largest in the United States with over 100 members. They regularly host competitions as well as an annual Roctoberfest in September. The mission of the club is to "enhance awareness and understanding of brewing."
https://www.centralarkansasfermenters.com/
08/07/23

11.2.6 What was your biggest batch as a homebrewer?

I was in the habit of making 20 gallon batches.

11.2.7 Did you ever brew a beer for an event (wedding, etc.)? How did it go?

Yeah, I would make beer to serve at community parties all the time.

11.2.8 Prior to getting into brewing what else did you do after graduating?

I got a job in a microbiology lab in Little Rock to determine if I wanted to pursue grad-uate school. I had a degree and wanted to use it, but I realized that working in a lab was very uninspiring to me. I treaded water for a while until I was convinced I wanted to pursue brewing, but there were no obvious opportunities around me at the time. My girlfriend at the time, now my wife, was here and I was hesitant to up and leave the area. Fortunately, Lost 40 opened at the time when I was getting desperate to leave the lab and move on. They opened and I managed to get a job as a brewer.

11.3 Current position

11.3.1 What are your current job responsibilities and tasks?

I direct all brewing operations and personnel.

11.3.2 What is your favorite and least favorite process to do and why?

The processes I find most fun are cultivating wild cultures and blending wild beers. My least favorite process is probably packaging, to be honest. It is very important for the business: if you can't package then you cannot sell, but packaging is not a process unique to crafting beer.

11.3.3 What classes that you took apply to some of the things that you have done in the brewing industry? Why/how?

Generally knowing how to work with SI units and basic lab wear is very practical for quality control, but microbiology is probably the most specific subject. Knowing how to bank and aseptically propagate pure cultures are very useful skills.

11.3.4 What adjustments did you need to make from working in a microbiology lab setting to a brewery setting?

The biggest thing is the physical component. You must be willing to use your body as well as your mind. In research, it's easy to imagine just sitting at a bench or desk for-ever. But brewing and manufacturing in general is much more physically demanding.

11.3.5 Do you have any advice for students wishing to apply their chemistry (STEM) degree toward getting into the brewing industry?

One of the coolest things about brewing is that it integrates a variety of STEM subjects. So, if you have a STEM background there are probably a variety of ways that you can make a significant contribution to the practice of brewing. So just identify what most interests you and dig in; the whole picture will unfold if you just keep following the breadcrumbs.

11.4 Experiences in the brewing industry

11.4.1 How would you describe your leadership style?

Listening and collaborating so everyone feels understood and impactful is very important. Not everyone is in the industry because they are passionate about the craft, but people can find interest and meaning in most anything if they feel supported.

Collaborative leadership
This was actually a common response from individuals in the brewing industry. Collaborative leaders are said to "often seek out different perspectives from their employees, peers, and higher-level leaders to have a full understanding of issues and possible solutions. Hierarchy isn't important to a collaborative leader, and employees from all levels are welcome to share their ideas, offer feedback, or jump in to help on a project."
https://www.fingerprintforsuccess.com/blog/collaborative-leadership-style
08/10/23

11.4.2 What are the various ways that you keep yourself grounded and you use to take care of yourself?

A cool thing about beer is that it's just beer. Aside from safety, the implications of beer are rarely serious, especially compared to other industries such as medicine, law, construction, natural resources and law enforcement. The worst days are the long ones, but that applies to every industry.

11.4.3 Have you had any negative experiences in the brewing industry that you would like to share?

No, not particularly.

11.4.4 Anything else you'd like to share about brewing in your particular area?

We have some awesome water here. It is very clean and naturally demineralized. It is great to be able brew with that kind of water.

11.5 Professional affiliations

11.5.1 What current professional organizations are you affiliated with? Why did you join them?

I am a member of BA, ASBC and MBAA for the educational and networking resources. Those organizations are fantastic.

11.5.2 What do you enjoy most about going to their events?

What's not to enjoy? Learning about what you love with people who love the same thing is the best.

11.5.3 How important is it to be associated with other professional organizations?

I think professional organizations are crucially important for developing and thriving in any industry.

11.6 Other/miscellaneous

11.6.1 Outside of brewing what other hobbies or interests do you have?

I play and listen to a fair amount of music. I exercise frequently. I am always reading about new subjects.

11.6.2 Is there anything else that you'd like to share about your personal or professional life that you think people might be interested in knowing?

Not really.

12 Jim Crute

BS, Biochemistry, Binghamton University, Binghamton, NY, 1980

MS, PhD, Biochemistry, University of Rochester, Rochester, NY, 1985

Post-Doctoral Fellow, Stanford, Biochemistry 1986–1990

Lightning Brewery, San Diego, CA, Owner/Operator Since 2004

Binghamton University https://www.binghamton. edu/ 08/30/23	University of Rochester https://www.rochester. edu/ 08/30/23	Stanford University https://www.stanford. edu/ 08/30/23	Lightning Brewery https:// www.lightningbrew.com/ welcome.html 08/30/23

12.1 College/degree pursuit

12.1.1 What inspired you to pursue a biochemistry (STEM) degree?

In biochemistry you are a jack of all trades and master of none. Although I did get to be pretty good at a couple of subjects, I am better at a variety of things rather than being a master of just a couple. I think for me it was that I had more options. It was not too specialized and I liked the work.

12.1.2 How long did it take you to graduate? Were there any stumbling blocks you'd like to share?

I was an undergraduate for 4 years and it was my goal to get the MS and PhD in 4 years – I made that by a month. I actually had a gap year where I went to medical school for a year and that was the biggest educational mistake I made. At least I figured it out right away. Other than that, I don't think I had any real stumbling blocks. As I look back, I did not know what I really liked until I took Biochemistry as a sophomore and decided that I would follow-up on that. Where I was it was one course less than a BA in Biology or Chemistry. I did not have any real issues with the advanced degrees except the typical graduate student stuff – am I doing the right thing or does my advisor really care? That is a pretty standard for graduate student experience, and I was not special in that regard.

https://doi.org/10.1515/9783110798777-012

Health Professions Advisory Committee (HPAC)
Many universities have an HPAC that is dedicated to helping pre-med students navigate
the admissions process and to assure that they are making the correct career decisions.
When I advise pre-med students I try to get them to shadow a physician or surgeon in
the medical field they are interested in as soon as possible. This not only helps them
with their application, but also gives them a small taste of what it is to be in the
profession.
https://www.uh.edu/pre-health/hpac/
08/30/23

12.1.3 Did you do any research as an undergraduate? If so, what was the project?

As an undergraduate I did do research. What I learned was largely was not to do in
regard to research. You need time, resources and access to the all the necessary tools.
I found there is just not enough time to do work as an undergraduate. Unless you do
work in a lab that has an on-going project where all that they need is a pair of hands.
The project I worked on was the recycling of tRNA between active and inactive tRNAs.
Looking back, we just did not have the money to do the needed assays in order to find
the enzymes that performed the processing. Now remember that was 45 years ago, so
it really was not a surprise. You learn that what you do needs to be practical and you
need to be able to connect the dots.

tRNA
This relatively recent article discusses tRNA recycling and is a relatively short read.
https://doi.org/10.1038/s41594-019-0222-1
08/30/23

12.1.4 Do you have any words of wisdom for students wishing to pursue a chemistry (STEM) degree?

The advice I would have is to do what you like because you'll likely be good at that.
Plus, be able to commit fully. This means that as much as you think you'll need to
commit, you will need to do significantly more.

12.1.5 How often do you use what you learned as an undergraduate in your current job?

I would say all the time. You know $PV = nRT$ (ideal gas law) and you need to understand
dissolving gases in solution. Another is the Bernoulli's principle of fluid flowing through

a tube. Then again, most brewers, either hobbyists or professionals, do not look at it like I do. I use those types of chemical principles all the time. People are usually surprised by the amount of chemistry that I use. Then I do find it a challenge at times to speak with people with an engineering background about chemical principles. They usually took their last chemistry course a long time ago. So I wind up battling re-educating them on reconceived prejudices rather than appreciating the science of it.

Bernoulli's principle
This is a great demonstration of Bernoulli's principle which relates the speed of a moving particle to its pressure – the higher the speed, the lower the pressure. It is important for brewers to understand this as it relates to moving wort or beer through lines, pump speed and pipe height. Other examples include sailing and planes needing to generate sufficient speed to take off.
https://sciencedemonstrations.fas.harvard.edu/presentations/bernoulli-beach-ball
08/30/23

12.2 Movement to brewing field

12.2.1 What was your "entry beer" into enjoying craft beer? Why did you like it?

I have to say Anchor Steam would be it though it might not technically have been classified as craft beer at the time (1985). I don't think it was classified as craft due to the volume requirements at the time. It had a great taste and was not too bitter or malty, but it did have some maltiness and bitterness to it. Plus, it was readily available and was consistent in quality.

Anchor Steam Beer
Many a beer drinker had their heart broken during the summer of 2023 when Anchor announced they were closing. Anchor Steam is a 4.9% ABV, easy drinking beer. It is described as being a "bright amber" with "toasted maltiness" and "velvety with lively bubbles." They put out a Christmas Ale every year that will also be missed.
https://www.anchorbrewing.com/our-beer/anchor-steam/
08/30/23

12.2.2 Did you get started with homebrewing?

I have been making beer since graduate school. Two of the beers that I make at the brewery now are similar to the ones that I made in graduate school.

12.2.3 What was your first recipe?

The first one was not drinkable. So, since I hate failure, I started reading books on homebrewing and beer styles. I thought ESB was a nice style of beer that I would enjoy. So that second recipe was a high-gravity ESB. That is one that I still make at the brewery. My friends from England tell me that what I made was actually a mid-East Coast English Bitters. I entered the beer into professional beer competitions and it won a number of awards. For example in 2012, the beer, "Red Arc IPA" was the best California Common style at the California State Fair. When you win a category award with a beer that is a good sign that you made a great beer.

12.2.4 Did you ever enter a recipe into a brewing contest?

Not so much as a homebrewer. I did with the brewery and the interesting thing I have to tell you is that it does not really drive sales. For instance, we also won the Best of Show award one year at the California State Fair for a Strong/Old Ale. Nothing good happened with sales the following year.

12.2.5 What was your biggest disaster?

When homebrewing, I was cleaning one of those glass carboys used for secondary fermentation one day. I would clean them out with bleach and rinse them out. Then, I would fill them with water and swirl them around to rinse them. The neck of one of the carboys nicked the corner of the sink and shattered into a million pieces. At first I thought I had not been cut and then I looked back down in the sink and realized that I had. I would call that a disaster. So, the next day I got rid of all of my glass carboys and switched to plastic immediately.

12.2.6 Were you part of a homebrew club? Which one? How long a member?

I actually was. I am a member of QUAFF in town in San Diego. Technically I am still a member but I guess it would be almost 20 years now.

QUAFF
This is a homebrew club in San Diego that holds annual competitions, monthly meetings and also recognizes a member each year who earns the highest number of points in periodic competitions that are member-only competitions.
https://quaff.org/
08/30/23

12.2.7 What was your biggest batch as a homebrewer?

I used to shoot for 10 gallons and I would end up with 8 gallons in the end. When you're doing that much you have to go a little slower which is good because you do not make as many mistakes as you would with a smaller batch.

12.2.8 Did you ever brew a beer for an event (wedding)? How did it go?

I did the beer for an Oktoberfest once for a friend of mine. That was an interesting story. We were out on the back porch drinking beer and he wanted me to do beer for his Oktoberfest. I asked him what beer that he wanted and he asked for a Kölsch, a Pilsner and a Hefeweizen. I said "I don't like Hefeweizens" and he told me yeah, but he knew I could do it. So I had this cooling refrigerator that I used for lagers and I did the Hefeweizen in there. It ended up tasting really good. I did get clove notes in there as opposed to the banana you get with the hotter temperature fermentation. I still do that one at the brewery as well and it won first place twice in the wheat beer category at the California state fair.

12.2.9 Prior to getting into brewing what else did you do after graduating?

I went to DNAX Research Institute in Palo Alto, CA, for a year and learned how to do cell culture and lymphokine purification work. I then went to the Biochemistry Department at Stanford University and worked on Herpes virus DNA replication and discovered several enzymes essential to Herpes replication. After that, I worked for a pharmaceutical company in Connecticut and worked some more on Herpesvirus DNA replication and found some inhibitors that could have been made into new drugs. Halfway through the management decided Herpes Simplex was a nuisance versus severe infection and cancelled the project. Then, I worked on signaling through the TNF receptor superfamily members and after that I worked on characterizing transcription factors involved in MHC transcription. I came out here, ran a gene expression department for a couple of years and found myself rearranged out of a job.

12.3 Current position

12.3.1 What are your current job responsibilities and tasks?

I am the main owner of Lightning Brewery. The brewery used to be bigger and distributed in Southern California. Then we had an issue with distribution 5 or 6 years ago. After burning through all of my cash reserves I down-sized to a nanobrewery where we just sell out of the brewery. So, my responsibilities are everything except sales for a couple of days a week.

12.3.2 What is your favorite and least favorite task to do and why?

Paperwork is a real challenge and is my least favorite. I do not have any emotional association with the paperwork and do not get any joy from doing it. But I still get it done. I do like making beer – that has a certain Zen quality about it.

12.3.3 What classes that you took apply to some of the things that you have done in the brewing industry? Why/how?

That is an interesting question. I would say intermediary metabolism is a big one. Basic Biochemistry applies because of how the pathways intersect. All the engineers think that you just want to take sugar and make it into alcohol. That may be true if you just want to make vodka, but beer has all of these particular flavors from the yeast – the esters and other compounds made out of intermediaries in the TCA cycle. Many do not understand that better wort is made when you step mash and it is pretty straightforward. I tell everyone this. You take 97 degree F water and add milled grain to let it sit for 10 min. That lets the malt solvate and allows the nucleic acids, in particular, to be broken down by endogenous nucleases. Among other products, this generates an acidic phosphate that slightly lowers the mash pH and that acid pH is closer to the optimal pH for subsequent enzyme reactions. Then I warm the mash up to a protein "rest" where proteases are more active and increase the amino acid content of the resulting wort. I then warm it up to the beta- and alpha-amylase mashing temperatures. In addition to breaking down starch to simple sugars it gives the grist the chance to liberate the celluloses, as in non-digestible carbohydrates, giving the wort more body and the beer more flavor since it does not sheet off the tongue as quickly when sipped. On the fermentation side, the yeast grows much better with the liberated free amino acids and nucleic acid in the worts. All of that is basic Biochemistry, but I don't think many folks will put 2 and 2 together when they are brewing.

Step mash brewing
Many readers are likely familiar with all-grain brewing and, similarly, many use a single-step infusion where they bring their grains up to a certain temperature (e.g., 152 °F) and hold for an hour or so to generate the fermentable wort. There is *nothing* wrong with this process, but step mashes where you utilize other temperatures and enzymes allow you to modify key aspects of your beer, including, head retention, clarity and mash pH.
https://byo.com/article/the-science-of-step-mashing/
08/30/23

12.3.4 What adjustments did you need to make from working in a lab setting to a brewery setting?

They were financial because you do not make much money at a brewery. When you manage people you try to be polite and let them know what you need them to do. If they do not follow directions again, then you may need to part ways. That is no different than biotech.

12.3.5 What do you enjoy most about your current position?

I can tell you this directly. It is that I do not have my parallel colleagues stabbing me in the back, my boss worried that I might want his or her job, or folks working for me saying things like "Jim doesn't like me." Much less unsettling.

12.3.6 Do you have any words of wisdom for students wishing to apply their chemistry (STEM) degree toward getting into the brewing industry?

My opinion is to keep it as a hobby and do not do it as a profession. It is not as rewarding nor as stable of a profession.

12.4 Experiences in the brewing industry

12.4.1 How would you describe your leadership style?

I'm pretty hands-off on the leadership stuff. I do not tell people what to do. I have and have had a variety people work for me and it would bother them when I would not tell them exactly what to do. That was fine, they could always ask me if they could not figure things out on their own.

12.4.2 What are some unique strengths that you think PhD biochemists might bring to the brewing industry?

I think that I am better able to understand what is actually going on when something is not quite right and fix it. It is important to remember that the beer trade is an execution business. People get rewarded based on doing the recipe and following those recipes to the letter. It is a great thing for people who do not know everything about brewing. Our motto is "Balancing the Art + Science of Beer." I like to think those beers were made better by changing one thing instead of several things at a time. I've always favored the saddle approach to a protocol. This is where protocols are setup so that some changes in execution have minimal changes in the outcome. This is in contrast to being on a peak and one step in the wrong direction causes you to fall off the mountain.

12.4.3 What are the various ways that you keep yourself grounded and you use to take care of yourself?

I hike 3 days a week and try to get a good sleep every night. I also try to keep a healthy weight and not drink too much. I am thankful to have some really nice friends and do fun things with them who enrich my life.

12.4.4 Have you had any negative experiences in the brewing industry that you would like to share?

I lost a business deal with a distributor who wanted a different beer. They put the beer that they wanted into the contract and dropped us in the process of making that change. With that I had to scramble for another distributor and burned through my cash reserves. My advice and learning experience was to keep as much control as possible and don't believe anyone else will act in your own best interests other than yourself.

Beer distribution
Beer distribution laws and regulation certainly vary between states and in many cases
localities. It is best to consult your local and state regulatory organizations about issues
such as direct shipping to consumers or self-distribution. Here is a site that has
this type of information compiled by state.
https://sovos.com/shipcompliant/resources/beer-distribution-rules-by-state/
08/30/23

12.4.5 Do you have any advice that you would give to your younger self before getting into the brewing industry?

Yeah, why don't you get a career in science and keep beer making as a hobby.

12.4.6 Describe your toughest day in the brewing industry so far

The day that the distributors dropped me, did not give me any reasons, and told me to get my stuff out of their warehouse in 30 days. Our contract said 60, so I did 59 days. Then they lost 200 of our kegs and I back-charged them for the lost kegs. Not a fun day, or actually days.

12.4.7 Anything else you'd like to share about brewing in the San Diego area?

So when I started Lightning there was 12 or so breweries in the county and there was a craft beer trade that had started 10–15 years prior. I thought we might have 30 breweries in the county in 10 years. Note at the time no breweries had gone out of business and I thought it was a pretty stable landscape. Instead, within 5 years there were actually 100 breweries, and within 10 years there were 200 breweries. So, there was so much brand dilution and confusion and it was hard to sell beer. And without selling beer you will not be brewing beer for long. Since then several of the largest breweries in San Diego County have either gone bankrupt or have been sold to mainstream breweries. Landscape makes a difference and here it is pretty complicated.

12.5 Professional affiliations

12.5.1 What current professional organizations are you affiliated with? Why did you join them?

I was a member of American Association for the Advancement of Science (AAAS) and American Society for Biochemistry and Molecular Biology (ASBMB). Those were expensive so I dropped them. On the beer side I was a member of the California Craft Brewers Association for 11 or 12 years. Then, I've been a member of the San Diego Brewer's Guild for 11 or 12 years.

American Society for Biochemistry and Molecular Biology (ASBMB)
This national organization covers both disciplines very well and offers members opportunities in "networking, professional development, meetings, journals and leadership." They hold an annual meeting that is typically in the middle of the spring.
https://www.asbmb.org/
08/30/23

12.5.2 What do you enjoy most about going to their events?

What is nice about the beer events is that the ones that I attend are actually the ones that make the beer. They are very positive and fun to talk to and have a beer with. There is a lot of camaraderie in that respect.

12.5.3 How important is it to be associated with other professional organizations?

It forces you to read the journal every week or every week. It is good to be forced to read stuff that you don't understand.

12.6 Other/miscellaneous

12.6.1 Outside of brewing what other hobbies or interests do you have?

I like music and regularly attend concerts in the area. I play the clarinet even if Covid put a stop to playing in a local wind ensemble. Hopefully I'll get back into that. I also do woodworking and if you call container gardening a hobby, I like keeping up with a varied herb garden. I did master the art of using sourdough during the Covid crisis for baking breads. I have had the same starter for 4 years now. And I have found it is a little like brewing in that you make small changes and generate small improvements that are reproducible. After a year of changes, what you get is at least as good as a standard bakery. My friends seem to like most of the things that I bake. Good thing since I would not want to eat it all myself.

12.6.2 Is there anything else that you'd like to share about your personal or professional life that you think people might be interested in knowing?

I think we covered it. Keep brewing as a hobby, and do what you are good at and like since you will certainly spend a lot of time at it.

13 Drew Herron

BS Biochemistry from St. Michaels College, Colchester, VT 2016

Brewery Group – Pure Madness (Roadhouse Brewing, Melvin Brewing), Quality Assurance Director

St. Michael's College
https://www.smcvt.edu/
08/10/23

Pure Madness Brewery Group
http://www.puremadnessgroup.com/
08/10/23

13.1 College/degree pursuit

13.1.1 What inspired you to pursue a chemistry (STEM) degree?

I always knew that I loved chemistry. I was fascinated by the prospect of organisms, humans, animals and microbes and the things that they can do chemically. At first, I thought that I wanted to do pharmaceutical research and I decided that wasn't the route for me due to all of the little things that did not seem morally aligned with both myself and my values. So, I focused on the chemical aspects and processes instead. I mostly focused on organisms, and when I was a junior I started to focus on the chemical processes associated with brewing and the microorganisms associated with that.

13.1.2 How long did it take you to graduate? Were there any stumbling blocks you'd like to share?

Four years. I have a minor in math and one thing that I almost broke down with was Calculus II. It was not an easy class for me. I front-loaded my college career which I'd suggest that a lot of people do. I ended up taking my core courses during my first two years. I worked very hard my first two years in college. A lot of people go straight into party mode which I did not. I got lucky and ended up taking college physics in high school which helped a lot and took quite a bit off my load in college. The math and math prerequisites for the degree can bog people down.

https://doi.org/10.1515/9783110798777-013

13.1.3 Did you do any research as an undergraduate? If so, what was the project?

Yes. So, during the summer of my junior year I did cholamine research. Cholamine is a protein tag and one of my professors did work on cholamine for his PhD. So, basically what he was trying to do was add a halogen group to the end of cholamine to increase the binding affinity of cholamine. That summer I did some of the research on figuring out how to get the halogen tag to attach and trying to see if it's beneficial. Unfortunately, I was a typical college student and did not end up finishing that project.

Cholamine

[^{15}N]-Cholamine is used in metabolomics research that allows researchers to tag metabolites through reactions with carboxylate groups. Scalable synthesis of halogen groups of this label is desired due to increased reactivity and detection capabilities. https://doi.org/10.1002/jlcr.3598
08/10/23

13.1.4 Do you have any words of wisdom for students wishing to pursue a chemistry (STEM) degree?

Yes. One piece of advice that I would offer was that my experience was much better because I went to a smaller school with a strong chemistry program. I got the opportunity to do research. I also TA'd in college which you will likely not get the chance to do at a larger institution. I had more one on one with faculty as a result doing a biochemistry degree there. Also, make sure you stick with it because I know a lot of students start off with the chemistry and biology and get frustrated with it so they drop out or change majors. I definitely found things more interesting as a junior and senior because I stuck with it.

13.1.5 How often do you use what you learned as an undergraduate in your current job?

Quite a bit. I probably use it every day. Much of the stuff that I use is on the biology side of my degree. I work in a lab and so I do the chemistry side as well with doing ABV, SRM and IBU analysis. I got my first job because of that cholamine research and the fact that I had GC experience. I got to go into a bigger lab and get more GC experience due to what I did as an undergrad. Biochemistry is the study of metabolism and brewing beer is literally yeast metabolism. So, having an understanding of yeast metabolism helps me every day in my job.

13.2 Movement to brewing field

13.2.1 What was your "entry beer" into enjoying craft beer? Why did you like it?

That is a tough question. I was living in Vermont. So, my entry beer would have to be Long Trails Double Bag which is a double alt beer. Long Trail later became the first place I worked in the brewing industry. I fell in love with dark beers and still love them. That made me want to drink more craft beer. I lived in Vermont at the time of the brewery boom which was when breweries started to become extremely popular. I drank a lot of IPAs, West Coast IPAs and then later Hazy IPAs. I think it was very popular in Vermont because that brewery was doing craft beer before it was even a thing. They started in 1989 and it was a thicker beer that had flavor to it. I was a college kid and it had a higher alcohol content which is something that college kids often think about.

Long Trail Double Bag Beer
Double Bag is described as a double amber ale with 7.2% ABV that is smooth and has roasted malt flavors as well as sweet notes including caramel and chocolate.
https://longtrail.com/beers/double-bag/
08/10/23

13.2.2 Did you get started with homebrewing?

I started homebrewing during my junior year in college. I and one of my friends tried to start a homebrew club on campus. We started an unofficial club that could not be affiliated with the college, we would brew at a house, anyone who wanted to join could come join, and it ended up being mostly me and him just doing the brewing and everyone else just partying around us.

13.2.3 What was your first recipe?

It was pretty bad. The first thing we did was an extract driven IPA or double IPA which was ambitious for a first beer which is also why it was so bad.

13.2.4 Did you ever enter a recipe into a brewing contest?

No. My homebrew career was approximately 15 beers. I realized that the stuff I made was not very good so it's kind of funny that I ended up in the brewing industry.

13.2.5 What was your biggest disaster?

One was definitely sanitation. I did not understand sanitation or anything like that at that time. I made extract beers and the way that we just mashed in extracts was not really good. The biggest issue was temperature control. The carboy was just at room temperature and so in Vermont you had big temperature fluctuations.

13.2.6 Were you part of a homebrew club? Which one? How long a member?

Just the informal one that I mentioned. The club was really just the two of us with people hanging around for a brew.

13.2.7 What was your biggest batch as a homebrewer?

Just the 5 gallons.

13.2.8 Did you ever brew a beer for an event (wedding)? How did it go?

No. Most of mine were just for my friends and I would just give them away.

13.2.9 Prior to getting into brewing what else did you do after graduating?

I graduated in May 2016 and started my first job in the brewing industry two days after that.

13.3 Current position

13.3.1 What are your current job responsibilities and tasks?

I am the quality assurance director for Roadhouse Brewing. I am also the only person in my department so I do everything. That is ABV, IBU, SRM and all of the microbiology which includes plating, polymerase chain reaction (PCR) and just sanitation programs which I've done for a few breweries. I also run a sensory training and quality program within the brewery. We taste all the beers that come down the line and make sure there are no flaws. I train people to look for flaws as well as what ingredients should taste like/can go together so that we can make the best beers possible. Currently one of the biggest projects is making our brewery paperless so that all of our data is in a database system. I am working on that now so that data analysis is automatic so if we see something wrong or different we just look into the data. The last thing and biggest part is to make sure all the fermentations go well. I work closely with my head brewer to figure out that the ingredients are okay, fermentations are going well and that our beers are within specs or if they need to be blended. We also look at what, if anything, went wrong and how we can fix it heading into the next time we brew that particular beer.

PCR

The PCR has revolutionized so many areas of our lives. It allows for small pieces of DNA to be amplified for further analysis. Areas PCR has impacted include genetic cloning, DNA sequencing, microbial detection, tissue typing and, of course forensics. The first article describing the technique was published in *Science* in 1985. PCR has been particularly helpful in the brewing industry for identifying microbial contamination in beer samples.

https://doi.org/10.1007/s10620-015-3747-0, Accessed August 10, 2023

13.3.2 What is your favorite and least favorite process to do and why?

My favorite process is sensory training. I really enjoy the teaching aspect of my job, and think that I can get people in the brewery excited about and get people motivated to actually do and become more adapted to understanding beer. My least favorite is when things go wrong and making the decision of what to do with the things that are wrong. By that I mean deciding whether a beer goes out on market or do we need to take other actions with this beer. Making those decisions is rewarding and nice, but it is still stressful to do.

13.3.3 What classes that you took apply to some of the things that you have done in the brewing industry? Why/how?

Microbiology for sure. Organic chemistry was and still is helpful. Most of my testing is general chemistry and uses general chemistry. Biochemistry was helpful like when we went over the Krebs cycle, and metabolism because most of our off-flavors and nutrients are all metabolism. If you can understand yeast metabolism you can make the fermentation process more efficient as well as make your beers more shelf stable. Those are the big ones.

13.3.4 What adjustments did you need to make from working in a lab setting to a brewery setting?

I think the biggest adjustment is making decisions on your own. Academically you are handed a test and they say to do it. As soon as you go to the brewery lab (or any lab for that matter) you're expected to make your own decisions and know when things need to be tested. You need to analyze the results that you have immediately rather than just seeing whether your results are like everyone else's.

13.3.5 What do you enjoy most about your current position?

I'll give you a couple of answers. First is the people that I work with. I work with very talented people who have been in the industry for a while and we work very well together. It's nice to be able to bounce ideas off people. I also love the way that microbes make flavors through fermentation. The scientific aspect of my job is great. I got really excited about sour beers about five years ago and I like analyzing how the flavors that microbes create and using the microbes to make those flavors. For example, I could make a wort and then add different microbes to end up with very different flavors out of every single one of them.

The same wort can taste different solely due to yeast. What?
Youbetcha. If you ever find yourself in San Diego, CA, a real treat is to visit the White Labs tasting room. There you will find several examples of beers that were fermented with different yeast strains. For example, an IPA fermented with WLP001 California Yeast will have a 7.1% ABV and grapefruit/pine/resinous tasting notes whereas the WLP008 East Coast Yeast generates a 6.8% ABV beer with orange zest/kiwi/lavender notes. Their website, in fact, has data on the different fermentation results for these side-by-side beers.
https://whitelabstastingroom.com/tap
08/10/23

13.3.6 What would you like to share from your experiences in sensory analysis and training people to do it?

I am going to backtrack a little bit actually. I worked as a technician before this job. I did it purposefully. I had other opportunities to go into management, but I wanted to learn as much as a technician as I could so I worked for some larger breweries first. This gave me the experience in beer, and running laboratory procedures that I would not have gotten if I became a manager. I was building sensory programs at these breweries, but it is harder to get people behind learning more about beer at a brewery where they are busy and don't have time. One of the best experiences working in a smaller brewery is that it is great to have people who are interested in learning more about beer. It is good to have a captive audience and to work with people that are passionate about beer. I have been in this industry a while and sometimes forget what an average beer drinker knows. It is great to backtrack and go back to the basic tastes and the differences between those and knowing how to taste beers. Even for our more experienced people in the industry it is good to review the basics and keep up with the training that people already have. I can see much better results because of it. Every year we have the chance to judge beer, and it's nice to sit at a table with many experts and know that everyone knows their stuff and is able to contribute to the judging and not be intimidated by other professionals at the table. Fortunately our brewery is also at the forefront of brewing technology which gives us the opportunity to run sensory on the new and upcoming hops, grain and yeasts in this industry. This gives us an even more unique sensory experience that you get at most breweries.

13.4 Experiences in the brewing industry

13.4.1 How would you describe your leadership style?

I think breweries need leadership and I approach it from the point of view that I am not the one that is always working with the equipment or on a daily basis. I have a lot of experience, but I also know they do as well. I take a lot of what everyone has to say and what they think when I put together SOPs and policies because they are the ones using the equipment. Everyone should lead in a similar way because if you just plow through and lead by that, people won't respect you as much (because you don't respect them) and you won't get what you need from your employees. Everyone's opinions do matter and when everyone is involved with quality then you end up with a better product as a result. Less things are missed and we have a policy that if someone has a problem we stop, shut down and be able to figure out what is wrong. That way we can fix it before we move on. I worked with many people who did things like that and make sure that everything is tied back to quality.

13.4.2 What are some unique strengths that you think biochemists bring to the brewing industry?

The brewing industry is not filled with many chemists in general so I think that one of the things is a knowledge of what is going on in your beer. It can be very helpful. Having said that, I have worked with some amazing people who do not even have a science degree. Some QA managers were never in a lab before they got into the industry. One of the big things that helped me was that I had worked in other labs before, so I know what I am doing with regard to measuring. When I was working in Vermont, we were doing benchtop trials for alcohol levels and we wanted to taste the beer and see if we could use our techniques to determine alcohol levels. I think I was 0.01% off based on a manual measurement vs the Alcolyzer measurement, and that goes back to being able to accurately read a meniscus. It also helps with knowing the deep chemistry that is going on within beer, and you can delve deep into the issues where somebody else without that background might not. For example, diacetyl is an off flavor in beer and it comes from yeast metabolism. It helps to understand that it is something to do with the metabolism and makes it easier to know how to fix that issue or just be patient enough to allow the issue to fix itself.

13.4.3 What are the various ways that you keep yourself grounded and you use to take care of yourself?

I am a big outdoorsman. I very much like to fish and ski which is great because of where I live. I live in Victor, Idaho and work in Jackson, Wyoming so I am right on the Tetons. That really helps to be able to get away. One of my first bosses had a great saying, "We work to live and do not live to work." That helps me because I can take a step back, go fishing for a day or skiing and that alleviates stress. Another thing that is great to alleviate stress is that I have great coworkers without it being stressful. I can also go home to my wife and 2 dogs and 3 cats and have a separate life and completely rant about my day. She has her wildlife degree and I can rant about science and she doesn't just zone out. She has learned way more about beer than she ever wanted to and I have learned more about wildlife then I have ever thought I would.

13.4.4 Have you had any negative experiences in the brewing industry that you would like to share?

Not all that negative. The only downside to this industry is it can be corporate at times. Some decisions that I would like to make get backsided for business reasons. Like keeping beer out in market too long when it is not the best quality it can be. Sometimes those decisions are monetary decisions and not ones that I would have made. That's not the case here fortunately and is the reason I love the company I work for now.

13.4.5 Do you have any advice that you would give to your younger self before getting into the brewing industry?

Yes, and it is probably surprising. I really wish that I had taken a coding or programming course in college. These days, I am basically learning coding right now particularly for data analysis. There are things like R and JavaScript and Query stuff that I wish I had taken rather than learning to build from scratch is very helpful. With all the technology out today it is smart to get experience in this field because it can make your life easier in any industry.

Query language (QL)
The QL program language is object oriented and used to provide an efficient mechanism to analyze information found in databases. This can be particularly helpful to brewers who do a good job of collecting information for each of the fermentation and brewing cycles.
https://codeql.github.com/docs/ql-language-reference/about-the-ql-language/
08/10/23

13.4.6 Describe your toughest day in the brewing industry so far

One day, me and my counterpart found a wild yeast in one of our beers. The biggest problem with wild yeast is that once it gets into your brewery, it can get into every tank and contaminate everything. So, we ended up spending the better part of 24 h testing and re-testing samples and being completely stressed out because we thought we had a wild yeast infection. Fortunately it was not that issue. I also had some stressful days dealing with wastewater. One day one of our brewers did not turn off a hose and overflowed our waste water system. I walked into three inches of waste water on the brewery floor and six inches elsewhere. It was not a fun day.

13.4.7 Anything else you'd like to share about brewing in Wyoming or your area?

Honestly, I think the biggest thing around here is that you have a lot of young breweries which is very cool and it is great to work for a brewery that is very talented and willing to share with some of the others around here. The brewing industry is very collaborative and it is great to be able to work with the Wyoming Guild and share and learn ideas. Living in the Tetons, Jackson has three breweries and on the Idaho side there are three breweries. It is great to have people around to help you get through. It is also great to be able to share ingredients back and forth. We can also ask one of the other breweries for help with this kind of stuff because they know that if that are in a pinch we will get them back at some point.

13.5 Professional affiliations

13.5.1 What current professional organizations are you affiliated with? Why did you join them?

I volunteer with the ASBC. I joined because I like the educational side of everything. I work with a taskforce on small breweries and making resources for small breweries. It is important to share and spread that knowledge. Recently, I will start working with BA on their quality committee and I want to spread the knowledge and give back instead of just learning things just sitting in on meetings. Another one is the Wyoming Craft Brewer's Guild on their education side of things. In the past, I never worked with, but gained a lot from MBAA. In Oregon, they helped with those educational things and it is the reason why I volunteer.

Wyoming Craft Brewer's Guild
The Wyoming Craft Brewer's Guild "exists to support and connect the Wyoming craft beer community through advocacy, education and promotion." At the date of access there were 29 member breweries. The Guild hosts an annual Craft Brewers Summit. https://www.wyocraftbrewersguild.com/ 08/10/23

13.5.2 What do you enjoy most about going to their events?

First is the seminars and it is nice to hear other's opinions and about all the cool things. The second is the collaborative aspect and I have met some of the best in the industry.

13.5.3 How important is it to be associated with other professional organizations?

In my opinion it is very important. The resources that they offer and the collaborative aspect really helps with learning and building confidence and just the resources and connections are amazing. Everything that I do in this lab, SOP-wise, etc has come from those organizations.

13.6 Other/miscellaneous

13.6.1 Outside of brewing what other hobbies or interests do you have?

Skiing and fishing are the two biggest ones. I also love to camp and hike. That is quite a bit of it. I am also a big animal advocate and that has a lot to do with my wife, but I was that way before though too.

13.6.2 Is there anything else that you'd like to share about your personal or professional life that you think people might be interested in knowing?

The one thing that does make me unique is that I have worked on both coasts and also in the middle area of the country. Most people in the industry are regional-bound having worked in just one area of the country.

14 Gary Spedding

BS Biological Sciences, University of East Anglia, Norwich, UK 1980

PhD, Biochemistry, Leicester University, Leicester, UK 1984

Brewing and Distilling Analytical Services, Lexington, KY, Owner

University of East Anglia
https://www.uea.ac.uk/
08/21/23

Leicester University
https://le.ac.uk/
08/21/23

Brewing and Distilling Analytical Services
https://bdastesting.com/
08/21/23

14.1 College/degree pursuit

14.1.1 What inspired you to pursue a chemistry (STEM) degree?

The biological sciences were something that I was interested in ever since high school. It was one of those classic stories that I got turned on to the field by one of my teachers. I was not doing well academically earlier on and then I was challenged with biology and then enjoyed chemistry and physics at that point in high school. I got excited by those topics. I think interestingly there were repercussions going back to that. I got excited by a science program called "Don't Ask Me." It was hosted in part by a famous fella by the name of Dr. Magnus Pyke who was featured on Thomas Dolby's "She Blinded Me with Science." He was very exciting and an energetic arm-waving nutritional scientist. He ended up working in the Scottish Whisky industry. I ended up reading some of his papers recently and thinking that he and a few of his colleagues on that TV show really got me interested in science.

Don't Ask Me
This was a really interesting TV show that undoubtedly inspired many scientists in England. It included demonstrations, responses to viewer questions and centered on science education.
https://www.imdb.com/title/tt0209789/
08/21/23

https://doi.org/10.1515/9783110798777-014

14.1.2 How long did it take you to graduate? Were there any stumbling blocks you'd like to share?

It took me 3 years with the BS and about 3½ years once I wrote the PhD dissertation and defended it. There weren't any real stumbling blocks to say. To be honest though, chemistry was not my strongest subject. I did okay in that and physics was actually my weakest subject. Biological sciences though was my strong suit. Through biochemistry I got more involved in the chemistry concepts. I would say that I had limited math training and did not get a strong grounding in advanced math in high school. So, if anything that has been a bit of a stumbling block for me. Though my English language teachers helped me prepare well, with advantages to my future writing activities.

14.1.3 Did you do any research as an undergraduate? If so, what was the project?

Yes I actually did. I did a little bit of research at the John Innes Research Centre which is associated with the University of East Anglia. There are a couple of Centers/Institutes down there in that part of the country. I worked on the cauliflower mosaic virus (CMV). I actually did dideoxynucleotide sequencing of the virus. That was fairly new technology then and, of course, is relatively common now and even superseded by other more advanced methods.

Cauliflower mosaic virus
Talk about a virus of firsts. CMV "was the first plant virus to be discovered to contain DNA instead of RNA as genetic material." Its genome was also the first to be entirely sequenced back in 1980. The DNA from this virus is still used to express selected genes in plants to help increase crop yield or other desirable characteristics.
https://doi.org/10.1046/j.1364-3703.2002.00136.x
08/21/23

14.1.4 Do you have any words of wisdom for students wishing to pursue a chemistry (STEM) degree?

I think it is a very exciting field. Many people nowadays get swayed into working with computers and technology. There is still a lot of excitement to be found in science. I encourage students to seek out programs that will teach them how to use and repair instruments. I gave a talk at an ACS regional meeting many years ago about this issue. We found that quite a few people in science today are more familiar with push-button technology. You know, they put the sample in one end and the results come out the other. They do not really know whether the results are valid or if something is messed up with the instrumentation. I encourage them to dig deep and really get into the nitty gritty of the instrumentation and understand what is going on with the processes that they are trying to solve.

One thing that I learned from my PhD advisor that may be of help to students is that I sometimes stalled on doing the research in the lab because I was too busy reading about where we come from. I was trying to learn how we knew about the structure of molecules. It got to the point where my advisor told me that I had to stop reading about science and actually do some. To stand on the shoulders of giants in order to understand that we know that a benzene ring is a benzene ring and that the electrons circling around the atoms are there. Students should not get too involved in the minute details or original sources of structures and trust that others solved the basics and the grounding concepts that they need to build upon. So, getting too detailed is not necessarily a good idea for them. More hands-on work encouraged rather than just book smarts/learning. For sure, learn the concepts but then get active and do something novel with that information. Take it to the next level.

Standing on the shoulder of giants
"If I have seen further it is by standing on the shoulders of Giants." This quote is often attributed to Sir Isaac Newton, but "can be traced to the French philosopher Bernard of Chartres . . ." https://doi.org/10.1007/978-1-4471-0051-5_5
08/21/23

14.1.5 How often do you use what you learned as an undergraduate in your current job?

Quite a bit actually. It has been part of the success of my business. I do not have that much time to do research and I have never been one to toot my own horn. I have a PhD, but I rarely go by Dr. I do consider that I have "great bench hands". That is something that many people actually do not have. I know when things are going right and if the instrument is working correctly and giving us correct results. The problem solving skills that I learned as a student are great for figuring things out up front. For example, if you have an assay that is suddenly not working, if you can identify or pinpoint one of the twenty pieces or components of the assay that is the likely cause of the issue instead of having to test every single one blindly to get things working again then you have saved yourself so much time. And have shown that you understand the facets of that assay or experiment. That in essence is "good bench hands."

14.2 Movement to the brewing field

14.2.1 What was your "entry beer" into enjoying craft beer? Why did you like it?

I had an early introduction to Real Ales in the United Kingdom when I did my first degree. I lived with a couple I had met in college. He was the librarian at that college and moved to Norwich where the University of East Anglia is located to pursue further studies of his own. When I say college there was a system when they raised the high school leaving age from 16 to 18 whereby not a lot of schools had the capacity to take on those students for the extra 2 years. So, in some towns and regions of the country they set up what are/were called "third form" colleges to take on those students. I then went to a different school – "College" for those 2 years, known then as taking the "A-level" (Advanced level) subjects. Basically, the advantages for those going to such colleges were you took just three subjects – beginning to specialize for a career. Quite often the teachers were well qualified and some with doctorate degrees. My subjects were biology, chemistry, and physics. As noted above, I got to know the librarian there quite well and I ended up living in the house he and his wife then owned in Norwich. He got me into Real Ales as opposed to traditional mass-produced lagers and I took a liking to those. Then I started to look at homebrewing early on and did research into it. That is what ultimately led to the position that I got at the Siebel Institute (the oldest Brewing School, in the United States – see below). I liked the Real Ales because they were fresh, vibrant and had so much flavor. At first, it was a learning curve. You know many people do not move from the sweet wines to the red wines because of the tannic flavor. It is the same thing with Real Ales that takes some getting used to.

Siebel Institute
The Siebel Institute has existed since 1872. They offer a variety of training courses, sensory training kits, and lab media. Their website even offers a handy Brewing Ruler that allows users to do various conversions using a slide rule setup.
https://www.siebelinstitute.com/
08/21/23

14.2.2 Did you get started with homebrewing? What was your first recipe?

Yes. I don't know about the first recipe. I do know of some more interesting ones though. One of the ones that I made was a Gruit which is an unhopped beer. It has been 30 years since I brewed it, and it is still in good shape – yes one or two bottles remain unopened! I used that recipe when I published in Zymurgy magazine – the publishing organ of the American Homebrewers Association. I got involved in homebrewing when undertaking my first post-doctoral research position in Guelph, Ontario Canada. Then my homebrewing days of "career" continued through my days of teaching at Butler University in Indianapolis.

Gruit
There is actually a website dedicated to gruit ales. A Gruit is a beer brewed with herbs as opposed to hops. Both hops and the herb blend do the same thing though – prevent beer from spoiling. If you like historical beers you will certainly enjoying reading about Gruit beers.
https://www.gruitale.com/
08/21/23

14.2.3 Did you ever enter a recipe into a brewing contest?

I actually did. I was not big into competitions back then though. I was just at the start of attending American Homebrewing Conferences where they had essentially attendee-driven internal competitive awards. The public enjoyed the beers and whatever ended up being the most popular won. That is different than what they do now at the GABF where they have the professional-level judging which I participated in frequently as a judge and sometimes educator pre-judging. One brew I made used a recipe from a group in England called the Durden Park Beer Club. They had a booklet where they presented their research into early English beer styles. In there they had an oatmeal stout recipe that called for so much oats that I realized I was going to have to sparge the viscous-sticky wort very carefully and put in some rice-hulls as a bed to help with sparging. I forgot to put them in at first and the sparge was a very, very slow drip-by-drip process. Anyway, we put them in at the risk of some oxidation. There was something like 8 pounds of oats in the mash. I would have loved to have met the judge and shaken his hand because the one comment that amused me most was that the beer was quite good but there were insufficient oats for the style! I might even have a bottle of that left but it probably is not in as good of a condition as the Gruit.

Durden Park Beer Club
According to this website, the club was founded in 1972 and is named after their first meeting location. You can actually buy the book that Dr. Spedding references above at their website (*Old British Beers and How to Make Them*)
https://beerandbrewing.com/dictionary/OBo1h9eqb0/
08/21/23

14.2.4 What was your biggest disaster?

The biggest disaster was my first all-grain brew. I was checking the temperature with a mercury thermometer and I smacked it on the side and it broke. That was probably my biggest disaster. I, of course, had a couple of not too bad, over-carbonated, bottle bombs and that is probably a common one.

14.2.5 Were you part of a homebrew club? Which one? How long a member?

I was a member of two clubs. One was called the Cross Street Irregulars. They were associated with Sisson's Restaurant in Baltimore, Maryland. Hugh Sisson ended up forming Clipper City Brewing (now Heavy Seas Brewing). One of the interesting things about that group was that fruit beers were not well known back then. So, Hugh had made a cherry stout and I asked what would have possessed him to make that. He looked around to make sure nobody else was listening and said that the stout did not turn out that well. I think that is the initial reason that IPAs became so popular. The heavier than traditional style use of hops was to mask some of the issues with the base beer. Though there is more to it than that. Earlier on, the first enthusiastic home-brewers-turned-commercial brewers were a little naïve and uneducated about more global styles. I once heard, at a Brew Pub Conference the chant – "we are all malt brewers, and we brew to style." Many classic beer styles do in fact have adjuncts as part of the grist formula – something the brewers avoided based on the mass-produced beers. A more scientifically and generally educated brewer is also still called for today.

I was also a member of the Foam Blowers of Indiana (FBI). One interesting story is that we used to meet in a wine bar in the basement. I was headed upstairs with another member and a fellow was rushing down the steps and asked, "Is this the Fraternal Order of Police meeting?" We told him that this is the FBI meeting and he hurriedly headed right back up the stairs.

14.2.6 What was your biggest batch as a homebrewer?

I just did the 5 gallon batches. I did not get into too many of the sophisticated pico systems or anything like that. Mostly metal and glass brew vessels. I did end up doing a production run with a brewer at one of the Rock Bottom Restaurants and helped brew a batch with them in Indianapolis.

14.2.7 Did you ever brew a beer for an event (wedding)? How did it go?

Yes, I actually brewed the beer for our wedding. Still have some of the bottles that I labeled up (though empty). I don't think I made spectacular beers. One of the reasons that I gave up homebrewing was that I got too busy with work. When I first moved to Canada I was disappointed with the available beers. Now, with the craft beer movement it is really quite a bit of a chore to do homebrewing, especially when there is so much good beer now available.

14.2.8 Prior to getting into brewing what else did you do after graduating?

I did my Postdoc in Canada and continued my work with ribosomes there. That did not pan out quite as well publication-wise. So, I then ended up with a position at Buffalo General Hospital. A student had been over to England and became my Best Man at my wedding. I met him back in England when I was doing my first degree and I remember he told me that his father worked there. When I was in Guelph doing my postdoc, I called down to the hospital and said if your name is Middleton and you have a son named Blackford Middleton would you mind passing the message down to him. I ended up meeting Dr. Elliott Middleton who was one of North America's leading allergy experts at Buffalo General Hospital. He had a crazy project that he wanted me to look at. He wanted to see if there were plant bioflavonoid synthesis genes in the mammalian genome. So he asked me how that could be done. It was something that I was starting to get into in Guelph with Southern and Northern blotting. I told him that we could look for those genes, but we did not get too far with that. We did, however, find a couple of plant flavonoids that inhibit the reverse transcriptase enzyme, so we published a couple of neat papers there. It is kind of interesting that now you even see this being tied into some of the Covid papers that are coming out. It turns out that those two flavonoids can be helpful in treating Covid because it uses the same enzyme that they inhibit by two different mechanisms. Interestingly, Dr. Middleton was a very astute guy. He believed that Linus Pauling was incorrect in regarding Vitamin C as the active principle against the common cold. He did not think that it was the Vitamin C that was the actual cure, but that it was sparing plant flavonoids by acting as an antioxidant. I do not know whether that was or has been proven formally as an explanation though.

Basically, I did a couple of postdoctoral positions after Buffalo and then to Southern Illinois, Carbondale (SIU). I did some work on trying to understand the protein synthesis capabilities of archaebacteria. I had turned down John Hopkins once to go to SIU but then ended up working there with David Draper who was a thermodynamic expert. He was involved in RNA work. During this spell I also wrote for and edited a major practical guidebook on Ribosomes and Protein Synthesis for Oxford University Press. After that I ended up at Butler University. I built the first Biochemistry lab for them in the chemistry department and continued the flavonoid work there. I had tenure approval all the way through, and a higher level administrator blocked it – we did not see eye to eye. He gave me the year off with pay and I had my research lab and continued my research and I also took a course at the Siebel Institute. I got excited about what they were teaching and wanted to open a brew store or brewery. I then offered my services to Siebel during my "sabbatical year" to teach Biochemistry. It was an interesting place to be. They had some small rooms with cots and facilities for visiting teachers – kind of like a small hotel area. The President then asked if I'd like to work for them full time and take over the labs. I took him up on that offer.

14.3 Current position

14.3.1 What are your current job responsibilities and tasks?

Now it is pretty much administrivia. I have a team that runs the lab. To continue the Siebel story, I worked there for a little over a year and then the campus closed down. They reorganized, and Alltech, which had started a process of acquiring them, which later fell through, hired the main team, and relocated us to Kentucky. So, I was working with Alltech when they had an auditor come in that told them they were trying to do too much. They were getting ready to move us off in different directions. I was disappointed as the lab I built there was just taking off. After getting to know me in Chicago, people were now sending in samples to me in Kentucky and I did not want to let them down with yet another lab closure. So, I started my own business. In 2002 I started my current company. I worked in a small lab in my basement and then in 2010 we moved to a well-established lab that might have once been owned by the University of Kentucky. We moved in there and have been there ever since. In 2019, a microbiologist working with me was moving to Denver and so we looked at each other and decided to setup a second location. Because of Covid we got off to a slow start there, and he is now making some headway there. We recently put in some new equipment to do some other work in-house there. A focus being on new nutritional mandates by the US government agencies.

14.3.2 Looks like you all do analytical services for the industry. What type of analyses do you do?

One of the things we say is that nothing surprises us anymore. My philosophy has always been that if it's testable then we will test it for you or find somebody who can. One of the more interesting cases we once had was a mouse in a beer bottle. We figured out that the mouse got in there on his own. It was not tampered with or a fraudulent case. That was certainly challenging. The first thing you look for, and this is pretty interesting, is that in most fraudulent cases they will look to see if the mouse has a broken neck. Why? Because that means that they grabbed a dead mouse from a mousetrap! The other thing they might look for is if the beer was pasteurized. If it was then they will look to see whether the flesh on the mouse was cooked or not. People are always curious about the mouse getting in after the beer had been bottled. Apparently, if a mouse can get its head into any hole then the rest of its body can as well. Another interesting project being our testing of powdered alcohol that was in the news awhile back. We tested it and it was, of course, a very interesting news item. We also took up the challenge when the Kombucha industry said no one could accurately measure alcohol in their beverages. We were a part of the committee established by the Association of Analytical Chemists (AOAC) to study this claim.

AOAC

The AOAC publishes an annual official methods of analysis. The 2023 edition does, in fact, include a method for determining Ethanol Content in Kombucha.

https://www.aoac.org/official-methods-of-analysis/

08/21/23

14.3.3 What is your favorite and least favorite part of the job?

I guess the administrative side is what I do not enjoy – it is more tedious. I enjoy going back into the lab when I have the time and doing some research. Now, the team can do most of the testing, but there are a couple that only I do. I enjoy doing those things and a little bit of research and UV–Visible spectroscopy. You know, higher end stuff.

14.3.4 What classes that you took apply to some of the things that you have done in the brewing industry? Why/how?

One of the things that I work on is understanding maturation in oak and so my understanding of thermodynamics and kinetics is important for that area. I am not the savviest at math due to my background. Enzyme technology is something that I was really good at. I think that really has helped me with small scale assays. I remember talking to a student at Butler. We were working on the plant flavonoid assays and the assay was 10 μL total volume. We got some new flavonoids that were sent from another researcher and they were too dilute. I remember the student shaking his head when I told them that we're going to need to use 3 μL of the sample and I said that was a phenomenal volume to use for the assay. The other thing that is interesting today from a scientific point of view is that I worked on ribosomes and I worked on a tiny part of it with 53 components and proteins and the like. One of the more fascinating things is you can take it apart, put it back together and it is still functional. Now I am working with beer on the macroscale and it's amazing to think about the small-scale aspect compared to that.

14.3.5 What adjustments did you need to make from working in a lab setting to an analytical setting?

I think my passion for research was one of the things that needed adjustment. It was not difficult for me to figure that out. I did IBU and sulfur dioxide tests at Siebel. I think the good bench hands and familiarity with methods helped. The ASBC Methods of Analysis was also extremely helpful. Having a good understanding of the equipment and those methods made setting the lab up straightforward. In the early days when you start a business, you want clients. It was tedious and I was working on a sample on Monday and then another one would trickle in. As it built up, I had 10 samples to do on a Monday. In regard to the business, it was a seat of the pants business. I did put together a business plan, but one of the things they say about that is that you put it together and don't look at it again. In principle writing one does guide you but sometimes you just need to make a quick (non-budgeted) decision to get a $26,000 piece of equipment. I did not take a salary for a few years and that may have been a mistake. One of the things that I learned from that is you need to pay yourself first.

Sulfur dioxide

The ASBC method for determining total sulfur dioxide in beer uses the p-rosaniline as a colorimetric reagent or a segmented flow analyzer for detection. The p-rosaniline method uses $HgCl_2$ and 0.1 N sulfuric acid is used as well. While several ASBC methods call for degassing beer, this particular one does not. The referenced ASBC document actually has a slide show demonstrating the entire colorimetric procedure.
https://www.asbcnet.org/Methods/Methods/Beer-21.pdf
08/21/23

14.3.6 Do you have any words of wisdom for students wishing to apply their chemistry (STEM) degree toward getting into the brewing industry?

Yeah, at one point it used to be that a brewer or distiller would take you on if you were willing to forego pay or take an entry level position. Going back several years ago, $20,000 was a good salary for a beginning brewer – actually not even a beginning brewer. One of the things that we learned is that if you are running a brewpub the rules and regulations are different. If you have food on board, then the brewers would get upset. Even though they are a brewpub, people were then really coming in for the food, so the chefs got paid more. My other advice is to seek feedback. One customer that I worked with had an issue with their product. It was accepted by some people, but rejected by others. It is important to consider what your reputation will be if you put out a defective product in the market place. It is sometimes better to just dump the product rather than try to off-load it on to the consumer. One other case in point was the guy that used to run the GABF judging competitions was able to convince his bosses to purchase a microscope when he was a head brewer. They wanted to know how much it would cost and he told them that it would be less than the cost of one lost batch of beer. Once he explained it in those terms they were fine with purchasing it. Another thing is that with sensory evaluation and quality control is to make sure you do due diligence on products going out into the marketplace. A scientific education is a must have there.

14.4 Experiences in the brewing industry

14.4.1 How would you describe your leadership style?

I do not think that I am necessarily the best business leader in the world. We have a team that helps to build things together. I call on the team to do it and that may be problematic when I start to think about retiring. It is more of a situation where it is me and the team as opposed to an employer–employee relationship. In dealing with a transition that can be an issue. My legal counsel still suggests that the top down approach is sometimes needed. I think anyone running a business needs to take some management courses. You need to treat people fairly and with respect. You do not need to employ somebody who does exactly what you do, you need to employ somebody who is doing something different and you need to let them run with it. At one time I worked for a company and it was micro-managed. When people were doing the same thing as the owner they ended up getting fired. That did not help the company and its growth potential. Employ qualified people that can do the job and then get out of their way and let them do that job!

14.4.2 What are some unique strengths that you think PhDs or academic folks bring to the brewing industry?

The strengths come back to offering more hands-on courses. One of the problems with universities today is turning out people who are book smart and can answer questions but still need more hands-on training in industry. They have lost the definition of a university. We need to think more about apprenticeship types of positions. The British system went away from this a few years ago. They used to have sandwich courses at the polytechnics. It was 2 years in with 1 year in industry and 1 year back at the college. Universities definitely need to look to business and these industries to get them direct training. There are some brewer training programs and businesses where you can go as a brewer and get some in class and at work experience. Universities that want to establish fermentation science or brewing/distilling programs need to think very carefully about this approach.

14.4.3 What are the various ways that you keep yourself grounded and you use to take care of yourself?

That is an interesting and challenging question. My wife calls me a workaholic. One thing for people in this business to keep in mind is that one could say that workaholism and alcoholism could go side by side, particularly with the product being readily available. I think you have to be cognizant on that. What I do for fun, if you consider it fun, is that I like to work with graphics design programs and am actually right now procrastinating on a report because I am doing some graphic design work. That is my relief at the end of the day. I do enjoy my work and am passionate about reading. I have done some white papers and presented freely on LinkedIn or in brewers and distiller magazines. It is no good if it's just for me. I could probably make more money writing articles. I don't think I'll ever retire completely though some people may suggest that I need to slow down a little. There is still so much for me to learn about this fascinating field I fell into. After standing on the shoulders of others.

14.4.4 Have you had any negative experiences in the brewing industry that you would like to share?

We had our catalog stolen from us word for word from a competitor. One of the interesting things is that some of those earlier analyses that I mentioned above that I alone did were listed in their catalog. It was a business that we were doing some third-party service work for. It was odd that we found out about it. It was a new person on board at their business and they kind of got offended when they asked a new BDAS team member for their test results. They were told that most results were ready, but some were not as we sent some samples out for affiliates to run for us. The customer asked what shop and our team member refused to tell them. Then our employee went to investigate their business website and looked at their new catalog. They saw that our catalog was being used word for word. We were offended by that and had to send a cease and desist. That was a definite negative. Another one is that you as a laboratory must be very careful about certifying a product as being safe. Many people ask for that certification so we have had some negatives from that perspective when all we can do is pass on the report results noting if in tolerance and legal allowable limits/specifications.

14.4.5 Do you have any advice that you would give to your younger self before getting into the brewing industry?

That is another tougher question. For me I think serendipity is a big part of where I am today. So, I think we already talked about taking a few business classes. Maybe I should have looked at people like Steve Jobs and others who did not get a full education and learn from them. You need to be passionate enough. Another piece of advice is to think carefully, consider your audience and realize that the product that you make or offer services on/for is not going to be something that everybody will love or need. Look for a more global education about running a business if you decide to build one. And where you can get some hands-on training or experience. Resources are available in most towns – seek out via the local library or government offices. Also, for the alcohol world find out who and all the authorities that regulate what it is you do. We did most of our due diligence but missed a few things that we were forgiven for, but it could have been worse.

14.5 Professional affiliations

14.5.1 What current professional organizations are you affiliated with? Why did you join them?

On the distilling side, the American Distilling Institute which was founded about the same time that my company was – Bill Owens founded that. American Craft Spirits Association was another one that we joined early on and they have grown as an agency. On the brewing side, we have ASBC and MBAA that have journals and information about production and methods of analysis manuals. We are also members of the IBD. We're involved with all of those for sure. I am also a member of the American Society of Enology and Viticulture but we do not do a lot of wine work so it is mainly looking at what is going on in the wine world as it relates to brewing and distilling. For example, maturation of wine in barrels relates to distilled spirit aging. We're also working on establishing a professional journal for distillers.

American Distilling Institute (ADI)
The ADI is "the oldest and largest trade association dedicated to craft distilling in the world" and was founded in 2003. Member benefits include access to auctions, an annual conference, educational materials and access to distilling based publications.
https://distilling.com/
08/21/23

14.5.2 What do you enjoy most about going to their events?

It's a little bit of a laugh around here. The team that I work with does not like to go across the trade show floor with me because they know we will not make it across to the other side. I either want to talk to somebody or they want to talk to me. I do not toot my own horn, but I recently had very influential people approach me at a meeting so that made me feel good. I have a good reputation on the distilling side and have missed some of the recent brewing ones due to being busy. I do encourage my team members to only ask genuine questions. A funny thing about one of these types of events had to do with the GABF pre-judging teaching we did. The crowd got to not trust us because we would always present something unusual or tricky to deal with. One year we did hop flavor additions to beers, and we knew that the aroma had changed, but not the bitterness and we had people who would come up to us and say how much more bitter the beers were. It is necessary to challenge people and it certainly is important to teach about sensory evaluation.

14.6 Other/miscellaneous

14.6.1 Outside of brewing what other hobbies or interests do you have?

Like I said earlier I like to work with Photoshop and do other graphic design work. I do enjoy growing tomatoes and peppers. I spend some time in the garden. We moved to a house that needed some work and had like 20–30 years of leaves in the yard that need to be taken care of. I like doing odd projects like building a new table. I do a few things that take me away from the lab though perhaps not enough.

14.6.2 Is there anything else that you'd like to share about your personal or professional life that you think people might be interested in knowing?

The serendipity thing. When something does not work out then build from it and try to figure out what you want to do and assess that. If you fall, pick yourself up and be passionate about what you are doing. Do not allow anyone to tell you that you cannot achieve your goals. Try not to repeat the failures of others. Also, family does matter, and I learned that as well. Make sure you make time for family.

15 Elizabeth Agosto

BS Biology, Minor in Chemistry, University of North Georgia, 2012

PhD, University of Pittsburgh, Molecular, Cellular and Developmental Biology 2019

La Cumbre Brewing Company, Albuquerque, NM, QA/QC, Lab Technician

University North Georgia
https://ung.edu/
08/09/23

University of Pittsburgh
https://www.pitt.edu/
08/09/23

La Cumbre Brewing
https://www.lacumbrebrewing.com/
08/09/23

15.1 College/degree pursuit

15.1.1 What inspired you to pursue a STEM degree?

I've been interested in science for as long as I can remember. My dad has always been a bit of a tinkerer. Growing up, we did a lot of stuff like building the homemade volcanoes or backyard rockets, you know, model rockets, stuff like that. We always had a telescope, so I was always looking at the sky or, you know, taking things apart, figuring out how to put them back together, so it's always been kind of "how stuff works" mentality that made me pursue science.

In regard to my graduate degree, I did a PhD at the University of Pittsburgh. I had 2 years of classwork, and during my second year I selected a mentor and did my comprehensive exams and things like that, and then from there did about 2½ years of research with my mentor Karen Arndt. Her work focuses on epigenetics and transcription control in yeast. It was what turned me on to yeast. My project was transcription control. It focused on transcription termination control. We looked at how histones – the proteins that organize DNA into chromatin in eukaryotic cells – form nucleosomes that serve as physical roadblocks to the transcription machinery and aid in transcription termination. If those nucleosomes don't come together like they should or if they're not modified with the proper histone modifiers the process can go wrong.

https://doi.org/10.1515/9783110798777-015

Histones
Histones are the backbone proteins for DNA and help with condensing it in the nucleus.
They also play an important role in modifying gene expression. When eight of them are
combined they form a nucleosome.
https://www.genome.gov/genetics-glossary/histone
08/09/23

15.1.2 How long did it take you to graduate? Were there any stumbling blocks you'd like to share?

It took me 4 years. Yeah, I mean my biology program was definitely challenging. It was hard making up my mind just because I think my interest in science going into it was so general. I actually started as a chemistry major and I think it was my second intro bio class and my professor was so passionate about what she was teaching and she was so obviously in love with the field. But I couldn't let go of chemistry, so I kept it as a minor. After taking physics classes, if I could have triple majored I probably would have. I'm sure a lot of students are finding that, you know, college is so expensive, especially if they're not willing to just stay with whatever in state, you know, institution they have nearby. I think that the practical side of following your passion is important too.

15.1.3 Did you have the opportunity to do any research projects as an undergraduate?

I did a small one. Our genetics professor in the department was really interested in taxonomy of some endangered fish in Mexico. It was something that he worked on, I think in his postdoctoral studies. I did some just like phylogenetic taxonomy research. It was pretty simple, looking back. It was just a lot of like DNA extraction, PCR, gene sequencing, but kind of got my foot in the door.

15.1.4 So do you have any words of wisdom for students that are trying to pursue a STEM degree?

As scary as those degrees may seem, the people that I've talked to have said, "Well I'm not really good at math or remembering everything is going to be an issue." Just do it anyway. Yeah math may be scary for you but there's always extra help and resources for those classes or small details that you are struggling with. There are so many ways, so many opportunities and resources like tutoring that are available. If it is something that you really love then don't let those fine details scare you.

15.1.5 How often do you use what you learned as an undergraduate in your current job?

Gosh, I mean all the time. I feel like really the basics of classes such as general chemistry, lab hygiene and basic lab math, dilutions for example and those are the things that I feel like I use the most. The basics and foundation of biology and chemistry that have helped me the most. I don't really use the work from my PhD that much, except for some specific knowledge of yeast metabolism.

15.2 Movement to brewing field

15.2.1 So everybody has an awakening when they have that first craft beer. What was your "entry beer" into enjoying craft beer? Why did you like it?

Blue Moon was my first craft-style beer. I was kind of boring, maybe, in college and really narrowly focused on my studies, so I didn't get out to parties or events that much. I got to graduate school and one of the ways that I used my spare time was to explore craft breweries. I went to school in Pittsburgh. Stuff was always happening at breweries, like bingo or trivia or live music. First, I kind of dabbled in like grocery store craft beer such as, you know, Fat Tire from New Belgium or Blue Moon. Then I started sampling these other really kind of whacky at the time sounding beers where they were doing these bingo and trivia events that I was going to. I thought wow, you know, I didn't know beer could taste like that. Blue Moon was so easy drinking. My dad would drink the "dad beer" like Coors. I'd have a sip or two and think, "Oh, that's kind of gross." That was horrible of me to say though now knowing now that those are very well done American lagers. Now, I appreciate the consistency associated with those, particularly from a quality standpoint. They are such a huge operation to consistently churn out the same product every time. Going back to Blue Moon – it was easy drinking, low alcohol and made of wheat which was totally different. It also had this beautiful orange aroma that was so different from anything that I had experienced at that point.

Blue Moon

Blue Moon is a 5.4% ABV Belgian White-style beer. It is described as a "wheat beer brewed with Valencia orange peels with a subtle sweetness and bright, citrus aroma." It is most often served with an orange peel slice that is noted for accentuating the citrus notes in the beer.

https://www.bluemoonbrewingcompany.com/en-US

08/08/23

15.2.2 Before you got into the brewing industry did you get started with homebrewing?

I did actually. During graduate school after hanging around these craft breweries I found out that homebrewing was such a big thing. It's kind of a silly story, but I was involved in this Improv comedy theater and they had me in to do various events with them. They had me in to do a "meet the scientist" show. One of my friends at the show called me the yeast whisperer. And I was like well I just use yeast in the lab, but from there I think that kind of got me started thinking about other yeast-centric hobbies and I got a homebrewing kit from Northern Brewer.

15.2.3 What was your first recipe?

It was an Amber from a Northern Brewer kit. It was a little 5 gallon recipe on the kitchen stove, extract beer.

15.2.4 Did you ever enter a recipe into a brewing contest?

I actually did. I did two competitions. I don't remember what one of them was called, but in Pittsburgh there was the Edgewood Symphony Orchestra that was doing a fundraising homebrew competition. I entered a Belgian Quad that did pretty well.

15.2.5 What was your biggest disaster?

My husband and I did a "clean out the freezer of hops IPA." It was not a good idea. We threw a lot of the hops that we had in as bittering hops. It was like this bitter and flavorless beer. We also had an epic boil-over in the kitchen which made a big mess.

15.2.6 Were you part of a homebrew club?

I was never part of a homebrew club. We moved to Albuquerque and never familiarized ourselves with the surrounding clubs. Now, we did move here at the end of Covid and there was not a lot of activity that we could see, but we never really reached out to them when things picked backup.

15.2.7 What was your biggest batch as a homebrewer?

We're actually at 10 gallons now. When we went to all grain we got a 10 gallon Stainless Steel system. We went with the larger system. It's almost a little too big now. Neither of us really drink that much and it turns out that 10 gallons is a lot, so we end up with oxidized beer.

15.2.8 Did you ever brew a beer for an event (wedding)? How did it go?

Nope. I have not done that yet, but it would be pretty cool to do it.

15.2.9 Prior to getting into brewing what else did you do after graduating?

Not much, honestly. After I graduated from Pitt it was kind of a weird time. I was pregnant with my first son and he was born in July of 2019, we moved from Pittsburgh, and then basically Covid was at its worst. I ended up staying at home with him for a year and a half. At that time it didn't make any sense for us to work with a child and no childcare. Then I lucked out and got the job with La Cumbre.

15.3 Current position

15.3.1 What are your current job responsibilities and tasks?

QA versus QC
If there ever existed two terms that are often confused, it would be quality assurance and quality control. One helpful way to distinguish these is to focus on the product itself. QC focuses on analysis of the final product to ensure it meets standards, while QA focuses on the process associated with generating that product to meet quality requirements. They are both part of the overall quality system. Another way to think about the differences is that QA is proactive and QC is reactive.
https://www.scilife.io/blog/qa-vs-qc-top-differences
08/08/23

I'm doing a good bit of the QA and QC. There are some things that I really don't have my hands on like packaging quality, some of the work with cans, consumables and kegs. Quality there kind of falls on the people who are doing packaging. On that side, I mostly just ensure the product going into package is what we expect, appropriate final gravity, pH, color, etc. I monitor dissolved oxygen and carbon dioxide in all of our canned beers. Of course, I'm checking cell counts in our packaged product to make sure we don't have too much dissolved yeast in suspension when things go out. I do handle the fermentation quality control. I am checking gravity and pH every day, doing microbial assessment by plating all of our packaged product and PCR on usually our most sensitive beers – anything that is low alcohol or low hop content. I am plating on different media to look for growth of wild yeast or spoilage bacteria. I crop yeast and monitor cell counts and viability to ensure yeast going into the next batch is good. Then there are some other odds and ends, depending on what we're making. I do a lot of distillation so if we need verification on ABV if we're worried that something might have over-attenuated, or if we're looking for vicinyl diketone content prior to crashing.

Vicinyl diketones (VDKs)
These are produced during the fermentation stage of beer production and can signal issues with fermentation. The two VDKs found in beer are 2,3-butanedione (diacetyl) and 2,3-pentanedione generating a buttery or butterscotch flavor, respectively. In fact, movie theater popcorn butter is sometimes used by beer judges to calibrate their palate for this off-flavor. The method for analyzing total VDK involves a distillation followed by reaction with α-naphthol and analysis on a spectrophotometer at 530 nm.
https://www.asbcnet.org/Methods/BeerMethods/Pages/Beer-25-MasterMethod.aspx
08/08/23

I am also looking to get a sensory program going to help come up with product descriptions for our main brands. At the end of the day I guess I can crunch numbers all day long, but sensory is where we really address our customers' perception of our products and their consistency. We are definitely trying to include people from all over the brewery and not just people on the brew side. You know like servers and distribution people just to have a variety of perspectives and more observations.

15.3.2 What is your favorite and least favorite analysis to do and why?

Being a microbiologist it might seem a little bad to say but I think filtering beer to do microbial analysis is not my favorite – it's kind of boring. Yeah it might sound odd to hear me say that plate analysis is boring. On the flip side I really enjoy doing the PCR work. We have the Chai Open qPCR system.

15.3.3 What classes that you took apply to some of the things that you have done in the brewing industry? Why/how?

I think the one for me has actually been analytical chemistry. You know being used to dealing with variability and making those technical measurements is still so very important. That is where I learned the most on being able to take good lab notes and being very organized.

15.3.4 What adjustments did you need to make from working in a lab setting to a brewery setting?

I think the good thing about working in an academic lab setting is that you have the ability to have such good control over the environment. Everything is very clean and tidy. Chemical disposal is already setup for you. When you are in an industry lab the focus is different. There isn't a larger, established institution that has everything in place for you. You are kind of flying by the seat of your pants and things are unpredictable. Every measurement comes with quite a few variables to consider. When you are working at such a large-scale things are very different from the microscale I got used to in grad school. Like our largest tank is 120 barrels.

15.3.5 What do you enjoy most about your current position?

When people think of the brewing industry they think that it's a fun job and I think that is primarily true. There is quite a bit of freedom in it. I find the environment to be supportive and our headbrewer actually has a background in biology. So, you know, my supervisors understand the importance of the numbers and the science. They let me try something different. Having a supportive team and some freedom is something that I really do appreciate about my job.

15.3.6 Do you have any words of wisdom for students wishing to apply their chemistry (STEM) degree toward getting into the brewing industry?

The major thing is it probably seems a little daunting because I think it's one of those things that you do not hear from a faculty advisor or something that pops up in conversations. If somebody wants to go into the field and is not getting a lot of advice or guidance the traditional way, then pop in to your local brewery and talk to them about it. The headbrewer or director of operations. Another general truth about the industry is that people are very friendly and welcoming. They are super happy and they like to hear that you are very interested in this.

15.4 Experiences in the brewing industry

15.4.1 How would you describe your leadership style?

I would say that I try to lead by example. I am a fairly Type A, meticulous person. I am really careful in my day to make sure that everything is clean and tidy and organized. That it is written down and accounted for. I leave careful notes for people that they say are easy to follow. I guess a good example is that I am training our head cellarman to take over the day-to-day in the lab. I try to be the example for him and I have showed him how to do everything my way, and then we discuss ways to change the protocol that make more sense for him. I am adaptive and flexible and understand that not everybody operates or learns the same way and try to come up with solutions to account for that while still reaching the same conclusions.

15.4.2 What are some unique strengths that you think women bring to the brewing industry?

That's a hard one. I think that men and women are certainly equally equipped in the field. Brewing has been a male dominated industry for sure. I do not think that is because men are better brewers, but it has just been the way that it is in most fields. But in general the modern woman has a different perspective, different life experiences that offer more variety to a growing field. It is good to have representation of various groups that gives a broader perspective, allows for influence from more varied experiences, to better reach a growing and diverse customer base.

15.4.3 Have you had any negative experiences in the brewing industry that you would like to share?

I do think that what we just brought up is an important thing. It can sometimes feel like a "boys club." Sometimes women feel left out, whether or not it's intentional. You know the guys have these interests in common or they have known each other for a long time. Sometimes as a women and a mom it is a different experience. You know, most of the time I can't hang out after work because I have to go pick up my kid from daycare, things like that. I don't think it affects my professional relationships with my coworkers so much as just the personal relationships, the camaraderie.

15.4.4 Do you have any advice that you would give to your younger self before getting into the brewing industry?

I feel like I could say this at any point in my life. Do more research and read more. The history of beer is so deep and there are so many different styles. La Cumbre is a fairly traditional-leaning brewery and brews a lot of traditional styles, as well as newer, popular ones. I do not know as much about the history and traditions that make some of these styles what they're supposed to be. The same can be said for staying on top of all of the new and innovative things that craft breweries are doing. I think you need to continue to make time to humor your own curiosity and stay informed.

15.4.5 Do you have any tips or tricks for maintaining a positive work-life balance?

I feel like maybe I am a classic academic. Somewhat of a workaholic. Prioritizing is very important. I think you need to realize that there are some things that can wait until tomorrow. I think I get some of the workaholic from my dad. He would always go into the office like six o'clock in the morning and then come dragging home at six o'clock at night. He's spent and he's tired. Growing up, my dad was the first person to tell me, "You know I wish I spent less time working and more time at home with you kids." As you get older you learn what is really important to you. Obviously your work is super important if it's something that you love but there are other things.

15.5 Professional affiliations

15.5.1 What current professional organizations are you affiliated with? Why did you join them?

I am currently a member of BA and the ASBC. I do not know if it's current but I think I'm also a member of the AHA, if not just in spirit right now. The research in both the Brewer's Association and ASBC is very helpful. The journals give me ideas on how to approach doing things. Helps me with the QC program at La Cumbre. I am still a conference junkie. I love going to those meetings and just seeing what research people are doing and what different experiences they are having at the brewery or in the lab. I think when you are just reading papers or watching seminars and things like that, it is easy for you to just distance yourself from what they are doing. The real research that you can apply to the lab. When you get down to it and sit down with somebody to talk to about this issue or how they are approaching it. They say like yeah they struggled with it and this is what they did or how they tried to approach the issue. I think it's the community aspect of beer. It is one of the best things about the beer industry that it is welcoming and how helpful everybody is.

15.5.2 What do you enjoy most about going to their events?

It is definitely talking to people. I'm somewhat of a textbook nerd and I like going to the talks and learning the latest and greatest in the brewing process and talking to them afterwards and getting their take on the issues or research they presented. It's always super helpful.

15.5.3 How important is it to be associated with other professional organizations?

I think it is very important. I did not have them on my radar in college. I've only been thinking about craft beer for the past 5–6 years. There has just been such tremendous growth in the industry in that time. So, it feels a little like the academic side of it is coming of age with new developments, recipes and styles. It can be overwhelming. I think the earlier you get involved the better you get. The webinars help and I don't know I haven't heard about too many internships in the brewing industry. There have been plenty of opportunities to talk to people and get involved in local beer festivals.

15.6 Other/miscellaneous

15.6.1 Outside of brewing what other hobbies or interests do you have?

I am a baker. I do love yeast pretty much in every capacity. I like making everything, bread, confections, cakes, pastries, you know. I've been making bread for my son a lot. My mom laughs at me. He has the fanciest PB&Js in day care. I do a lot of baking, but I also like gardening, hiking, stargazing. I feel like it's what all of my bio friends love, making things and being immersed in nature.

15.6.2 Is there anything else that you'd like to share about your personal or professional life that you think people might be interested in knowing?

These types of questions are really hard. I don't think of myself as a terribly interesting person. I don't know. Actually, I taught my husband how to homebrew. I think it's how I won him over. He had the first beer that I ever made, that Amber. He thought it was pretty cool that I was serving a beer that I made. So yeah I taught him how to homebrew. It is something that we do together now. Being a software engineer, he put our new system together. You know, the engineering and gadgets are what he likes most. I like being a geek. People should not be afraid to share their goofy passions. You might inspire somebody to do what you love and then you have somebody else to talk to about it.

Section III: **The biologists**

16 Kent Ball

University of Iowa, 2002, Biology

Big Grove Brewery, QA/QC Director

University of Iowa
https://uiowa.edu/
08/22/23

Big Grove Brewer
https://biggrove.com/
08/22/23

16.1 College/degree pursuit

16.1.1 What inspired you to pursue a STEM degree?

I was a science geek growing up. I went to school to be a doctor and that did not really pan out because I did not want to spend eight years in school. I was originally pursuing a chemistry degree and did not really understand physical chemistry so I switched to biology and that was what I got my degree in.

16.1.2 How long did it take you to graduate? Were there any stumbling blocks you'd like to share?

It took me five years. In school, I was not very good at calculus and that was part of what prevented me from getting a chemistry degree. It was my ability to do higher level math and apply that toward quantum mechanics and the theoretical chemistry was an issue.

16.1.3 Did you do any research as an undergraduate? If so, what was the project?

I actually did not. I did work as a prep room chemist for the department at the university though. I was an undergraduate chemistry TA and I would help setup and tear down the labs for our organic chemistry labs.

https://doi.org/10.1515/9783110798777-016

16.1.4 Do you have any words of wisdom for students wishing to pursue a STEM degree?

Brush up on your math, because many of those fields are very math heavy. All of that math stuff that you think you do not need to know in junior high or high school. I use algebra and spreadsheets every day. You need to learn how to make data work for you because that is a skill that I use every day. Data is great, but if you do not know how to use it then it does not do you any good.

16.1.5 How often do you use what you learned as an undergraduate in your current job?

I'd say very frequently. I took several biochemistry courses and brewing is just applied biochemistry. You have enzymatic reactions and water chemistry. Microbiology definitely plays into it and we even do quite a bit of PCR. On the analytical side of things we do a lot of chemical testing.

16.2 Movement to brewing field

16.2.1 What was your "entry beer" into enjoying craft beer? Why did you like it?

It was Schell's Firebrick. It is an amber dark lager from Schell. A friend had a keg of it at his graduation. The first time I had it I could not believe it was beer. It tasted fantastic and had me wondering what I had been drinking for the past few years. I liked the fact that it had flavor and balance and maltiness. It was roasty but not over the top. I had not experienced something like that before with the hops and all.

Schell's Firebrick
Schell's Firebrick is a 4.8% Vienna Style Amber Lager. It is described as being a "refreshing Vienna-style amber lager . . . with a hint of hops combined with a subtly maltiness."
https://www.schellsbrewery.com/our-beers/firebrick/
8/22/23

16.2.2 Did you get started with homebrewing? What was your first recipe?

I actually did. I bought my dad a kit for Christmas. He had shown an interest in it and then we went to the basement – had a little stove down there. Mom then told us that we could not brew in the house ever again. After that we went out and got one of those turkey fryer burners and that was what we brewed with after that. It was a brown ale that came with the kit.

16.2.3 Did you ever enter a recipe into a brewing contest?

I have entered a few. I had mixed luck with contests and became a judge. Then I learned how the competition process works. I entered a small brew that I did not take care of at all. I brewed it, put it in the fermenter and went on vacation for ten days. I just let it ferment in the basement. When I got back I racked it into a keg and took it to a competition. I ended up getting best in show. It was a German Heffe beer. It worked out really well. I did not do anything special to it and it ended up working out very well and was the best homebrew I ever made.

German Hefeweizen
The 2021 BJCP Guidelines has this classified as a Weissbier. The ABV range for this style is 4.3–5.6%. There are more banana and clove notes in this style as compared to an American wheat beer. The overall impression of the beer is that it is a "pale, refreshing, lightly-hopped German wheat beer with high carbonation, dry finish, fluffy mouthfeel, and a distinctive banana-and-clove weizen yeast fermentation profile."
https://www.bjcp.org/style/2021/10/
08/22/23

16.2.4 What was your biggest disaster?

When I went into all-grain I just had one bucket. I had iodine as a sanitizer. I milled into that bucket and put the milled grain into my brew kettle. Then I used that same bucket to make my sanitizer solution using that to sanitize my equipment. What I did not realize was that when iodine and starch come together they bind up and that made the iodine completely useless. I ended up inoculating all of my equipment with lactobacillus. I ended up making my first sour that day. Made a really good Berlinerweisse but that was before I really knew anything about sour beers. I did that twice and then realized what I was doing. You know that starch test that you did in college? Well that was what I was basically doing with the iodine and the remainder of the milled grains. I don't know how many people still use iodine in homebrewing anymore. I grew up on a dairy farm and iodine was our primary means of sanitizing things.

Sanitizers
There are a number of sanitizer options available including iodine-based, acid-based, chlorine dioxide and even bleach. Many of the commercially available sanitizing products do not require rinsing, while bleach, of course, does require rinsing. All of them work very well, though some, such as iodine-based, can stain brewing equipment.
https://fivestarchemicals.com/products/sanitizers
08/22/23

16.2.5 Were you part of a homebrew club? Which one? How long a member?

I was actually part of one. I was part of Cedar Rapids Homebrew Club for several years.

Cedar Rapids Homebrew Club
Now known as the Cedar Rapids Brewing Society, this club hosts several events including Happy Hours and Beer Festivals. They emphasize the importance of drinking local beer.
https://cedarrapidsbrewingsociety.com/
08/22/23

16.2.6 What was your biggest batch as a homebrewer?

Ten gallons was my biggest batch.

16.2.7 Did you ever brew a beer for an event (wedding)?

I did brew the beer for my sister's wedding and my brother's wedding.

16.2.8 Prior to getting into brewing what else did you do after graduating?

I worked for a biotech company here in town as a production scientist and also took care of the materials QC. I tested all of the materials used to produce oligonucleotides – short DNA chains. From there I went to the university and did some research in the pharmaceutical field both in the lab and in the clinic. Once our grant dried up I went to the community college and went back to the prep room and ran the prep room at a community college here in town.

16.3 Current position

16.3.1 What are your current job responsibilities and tasks?

Basically I am the QA and QC person for the company. My job in a nutshell is that. We make 24,000 barrels of liquid per year. I take part in the development of procedures, I audit all of the logs every day and make sure that procedures were followed and if they weren't, then why weren't they followed. I help to improve those processes. I monitor the fermentation from beginning to end. At the end of the process before it is packaged I run a battery of tests and make sure that everything is up to our standards.

16.3.2 What is your favorite and least favorite process to do and why?

My favorite process is identifying problems and helping people fix them. When we have an issue and even if we don't know what the issue exactly is then we help track it down and narrow it down. My least favorite process is when we do have an issue that we do not know why it came up and it has come up too late in the process. I don't like the helpless part of that. You know it's not right but you don't know why it's not right. By the time you get to that point sometimes it is too far gone to actually be able to help out with the problem and provide meaningful input.

16.3.3 What classes that you took apply to some of the things that you have done in the brewing industry? Why/how?

Statistics for sure. Honestly, some of the calculus that I took that I stumbled through applies too. Microbiology and biochemistry are two other classes that I find myself using on a regular basis. Organic chemistry is another one and all of the lab classes that I took. At Iowa we had basic measurements classes where you get into a lab and you do things so that you can replicate them. Analytical chemistry was another one that comes into play. Actually high school algebra is used quite often in brewing.

16.3.4 What adjustments did you need to make from working in a lab setting to a brewery setting?

The biggest difference is that I came from a controlled lab environment where everything was document controlled by the FDA and learning to loosen that mentality to support innovation was a big adjustment. At the end of the day it is just beer as opposed to the chemodrugs that I used to work with. It is liquid happiness at the end of the day.

16.3.5 What do you enjoy most about your current position?

The industry is constantly changing so it is both fruitful and maddening. I got into the industry eight years ago and it was a completely different playing field at that point. There are so many more styles and so many new styles that are coming out. Fruited sours and hazy IPAs were not even in existence back then. They were still novelties at that point. The landscape is so unique now compared to that. Hops are always very innovative nowadays. There are so many new products coming out like hop oils and hop extracts that are constantly coming out. Environmental factors are coming into play. It is fun to see what the people on the other side of the supply chain are coming up with.

16.3.6 Do you have any words of wisdom for students wishing to apply their chemistry (STEM) degree toward getting into the brewing industry?

I would strongly recommend taking some food science courses. That seems to be what people in my position at other companies are holding in high regard.

16.4 Experiences in the brewing industry

16.4.1 How would you describe your leadership style?

I would say it is pretty blunt and to the point to be perfectly honest. I do my best to lead by example. I show you how to do something and expect you to do it that way. The best part of this industry is at the end of the day you can sit down, have a pint together after the shift, and still be friends.

16.4.2 What are the various ways that you keep yourself grounded and you use to take care of yourself?

It is very hard to get too far off the ground when you have things like Untappd telling you everything that you're doing wrong with your beers. I mean at the end of the day it's just beer. It is the working person's beverage and it ties everyone together regardless of your class. That is the beautiful part of it right? There is a beer for everybody who comes through our tap room everyday.

16.4.3 Have you had any negative experiences in the brewing industry that you would like to share?

Yeah. We had some products that snuck by that should not have been released. There were times that we that we thought the typical consumer would not pickup on the flaw. Lo and behold they did and they let you know about it a week later. Product recalls are never fun. That in a nutshell is never a good day. It is not just your reputation or anything. It is literally money being poured down the drain.

16.4.4 Do you have any advice that you would give to your younger self before getting into the brewing industry?

I would get into it at a younger age. That is for sure. I got into it at 35 and it is a young person's game. I am in my 40s now and the younger ones are definitely doing fine. Be flexible and willing to learn things.

16.4.5 Describe your toughest day in the brewing industry so far

I really do not know. Every day is a different challenge and you just have to get through it.

16.4.6 Anything else you'd like to share about brewing in your area?

Being in the Midwest and Iowa in particular is interesting. We are in the heart of Busch light country here. So, converting people to craft beer is a challenge. We definitely cannot beat that price point, but if we put out a good product then people might actually come over to liking our stuff. I mean, like my grandfather drinks Old Milwaukee ever since I've been around. I will bring him some of our beer and he'll drink it and he'll actually enjoy it.

Busch light
The 2023 Busch commercials keep ringing through my skull as I write this. Busch light is a 4.1% ABV beer with 95 calories and 3.2 carbs. It is "brewed longer to create a lighter body and fewer calories while always delivering that classic taste."
https://www.busch.com/products/busch-light
08/22/23

16.5 Professional affiliations

16.5.1 What current professional organizations are you affiliated with? Why did you join them?

I am a BJCP certified judge. I joined that to learn more about beers and different styles. We have an institutional membership to ASBC which I use their methods for most of my testing. I also use the ASBC journal to see what the latest research is in the field. MBAA is another one due to the constant research that is going on which is great. They have some cutting edge techniques that they offer in their technical quarterly. I am a member of the Quality Subcommittee for BA. I have attended two meetings and enjoy being around like-minded people. I am one of one in a company of a hundred who knows what I do and how to do it. I like being around people that I can ask for help and that are of legendary status in the industry. I am a member of the education committee for the brewer's guild.

Quality Subcommittee of BA
The Quality Subcommittee offers a number of excellent resources for BA brewers ranging from monitoring beer freshness to managing off flavors to tracking programs. The resources are available to BA brewers. As stated on their website, the Committee "envisions a membership that consistently produces beer of high quality."
https://www.brewersassociation.org/author/quality-subcommittee/
08/22/23

16.5.2 What do you enjoy most about going to their events?

Again, just the networking and being around people who view beer the same way that I do. It is great for me to be around people who understand the technical side of beer and appreciate how little changes can have huge effects.

16.5.3 How important is it to be associated with other professional organizations?

I would say to get into the industry first but once you are in the industry you can benefit from having people to ask questions about things. I dedicate maybe 10% of my time trying to learn stuff and be on the cutting edge.

16.6 Other/miscellaneous

16.6.1 Outside of brewing what other hobbies or interests do you have?

I enjoy fishing and coach archery as well as baseball at the local high school. Cooking is another one. I hang out with the chef's around here and we talk about techniques and am constantly inspired to go home and try new things in the kitchen

16.6.2 Is there anything else that you'd like to share about your personal or professional life that you think people might be interested in knowing?

I did not see myself working in a brewery when I was younger. It has definitely been a curve ball and if you can hit it that's a rewarding experience.

17 Adam Fleck

Portland State University, 2012; Environmental Science

Alliteration Ales, Boise ID, Production Manager

Portland State University
https://www.pdx.edu/
08/31/23

Alliteration Ales
https://drinkalliterationales.com/
08/31/23

17.1 College/degree pursuit

17.1.1 What inspired you to pursue an environmental science degree?

I have always had a high science and math aptitude in general. I was curious growing up. I liked environmental science at Portland State because of the broad applications that they had built into the program. They had a university studies program which was designed to help round out a student's skills. So, instead of just studying environmental science and doing tree cover studies or biological density analyses they really wanted to expand your expertise into other areas. I took social science courses and philosophy courses that they offered. It was also through electives that were required. So, if I was required to take a science class that was outside of my degree I would take biochemistry and geology. There was an ethnosociology course that I also took. The degree was broad ranged and I also knew that any degree with an environmental background in it would be marketable.

Ethnosociology
This is an interesting field of sociology that focuses on different ethnic groups. The author of the cited book is said to "offer a profound philosophical approach to the categories of the 'ethnos, 'narod,' 'nation,' and 'society.'"
https://www.goodreads.com/book/show/51176896-ethnosociology
08/31/23

https://doi.org/10.1515/9783110798777-017

17.1.2 How long did it take you to graduate? Were there any stumbling blocks you'd like to share?

It took me 6 years. The big one that I had when I first went into college I had a different major and ended up changing majors from computer science. I was into computers in high school and thought the degree was marketable. I knew that I wanted to get a degree in something I could use after college. For example, I actually have Calculus III as an elective which not many people with an environmental science degree would do willingly. One of the other things that was more formative for me getting into the brewing industry was some stuff that I did outside of the classroom. College is not just good for the classroom experience, it is good for expanding your horizons. I ended up doing beer and liquor customer samplings for a promotional company. I would go into liquor stores and ask people if they wanted to try a sample of a whiskey or a new beer. I was doing that for extra money. That was my first experience of the industry side of the field. I ended up working for that company for a little bit.

17.1.3 Did you do any research as an undergraduate? If so, what was the project?

I do not believe that I was part of any formal research. We had some of the postdocs and TAs in our lab courses, but I wasn't part of a formal research group or anything.

17.1.4 Do you have any words of wisdom for students wishing to pursue a chemistry (STEM) degree?

From that standpoint and this would be more from knowledge while I was going to college. I would have to say to work on relationship building. Knowing your professors better and not just knowing them, but knowing those who had connections in the field. My computer science professor spoke a number of times that she had worked for HP and Intel and some of those big companies before she retired and worked in the industry. Her knowledge might have been outdated and she did mention that. She was a very good professor though and did have that industry-based advice. With STEM degrees, you need practical application of your skills and not just the education from the classroom.

17.1.5 How often do you use what you learned as an undergraduate in your current job?

Quite a bit now. We are just a three-barrel nano system in Boise and just got started last September. Me and my significant other work there. I do the lab work and she does the marketing and the creative side. I do the beer production and recipe design side. I would say that what I have used the most was a lot of stuff that I learned outside of the classroom. I was one class away from a business minor, but I do find myself using what I learned in my business classes. Those have helped immensely in regard to starting a business.

17.2 Movement to brewing field

17.2.1 What was your "entry beer" into enjoying craft beer? Why did you like it?

The first one that I bought when I was 21 was McTarnahan's. They don't make it anymore and it was Portland Brewing Company's McTarnahan's Amber Ale. They have not been around for a few years. I don't remember what I liked about it. Even when I was going to college parties I preferred the quality over quantity. I remember that amber and red ales were my first foray into craft beer. My parents always had Henry Weinhard in the house which was a Portland beer staple for like 100 years.

McTarnahan's Amber Ale
As Adam mentioned, this one is no longer available. A review of this 5.1% ABV beer describes the beer as having a "effusively fruity aroma, with piney, sweet-orange hops and caramel-tinged malt." The brewery described the beer as a "copper-hued amber ale dry-hopped with Cascades from the Northwest. Bright, crisp, and complex."
https://beerandbrewing.com/review/portland-brewing-mactarnahan's-1605214445/
08/31/23

17.2.2 Did you get started with homebrewing?

I actually did. I got started toward the end of college. Very similar to other people, I was broke in college and wanted craft beer. I had a buddy who was going to brew beer and he invited me over to brew and so I was excited to go. With my science background I wanted to know how things work and how beer was made. It's a unique blend of science and alchemy. That is really what we do in our brewery now and I like to play off of that continued interest.

17.2.3 What was your first recipe?

It's been too long to remember exactly, but it was a partial mash for sure. Probably a Rogue Amber ale clone or some variation of it. Even starting out, we weren't following guidelines very closely.

17.2.4 Did you ever enter a recipe into a brewing contest?

I actually did. I like to say that it got the courtesy 35/50. I have a beef against competitions. It's a little like battle of bands judging to an extent. Some of the judges are lo-
cals who are judging and did not ever make it big. So, they may have a bias or want to
set the tone for the judging. Back to the beer, one of the things that we did was we
tried to make a floral and fruity IPA with late addition hops and dry hopping. The
problem was that when I brewed I messed up my calculations – had been brewing for
a year or so. I ended up with an 8.5% beer instead of a 4.5% beer. It was supposed to
be a session IPA. I entered it in the Imperial IPA category but gave it the title "Session
IPA." All of the scores indicated that the beer was too alcoholic, too dark and too bitter. It had too much of a fruity nose. So I think it was judged at least partially based
on the name.

17.2.5 What was your biggest disaster?

Bottle bombs were my biggest disaster, Bottle bombs, but in 22oz glass bottles. They
shot glass through the cardboard box they were conditioning in.

Bottle bombs
Most homebrewers who started off bottling have experienced this at one time or
another. The garage of our first house was victim to a batch of root beer we failed to put
in the refrigerator after it was finished. Common reasons for these include using too
much priming sugar, infections, bottling too early or using the wrong bottle.
https://foodbeerstuff.com/6-mistakes-that-lead-to-bottle-bombs-how-to-avoidfix-them-
home-brewing/
08/31/23

17.2.6 Were you part of a homebrew club? Which one? How long a member?

I was not big into the homebrew scene during college. Portland has a huge number of
homebrew clubs and, in fact, one of the oldest ones in the country. I did get into it
once we moved to Idaho.

17.2.7 What was your biggest batch as a homebrewer?

I did a couple of 10 gallon batches, All-grain.

17.2.8 Did you ever brew a beer for an event (wedding)? How did it go?

Yes, I did a couple of weddings and specific parties that had interesting recipes. I even did an IPA for a friend's wedding – also did a cider and a commercial lager.

17.2.9 Prior to getting into brewing what else did you do after graduating?

What got me into the industry was the oil field. I was working for this promotional company and ended up getting hired as one of their training managers. We would send people information on the products before they would go out and give out samples. Then a buddy of mine was in the oil field and with my science background we ended up doing on-site geology for oil rigs. That is where I was trained in GC. We were doing continuous monitoring using GC-FID for natural gas coming out of the wells. Of course, we were using microscopes and doing staining too. That was my first time using GC and, of course, being on the oil rig you are far from everything. If the equipment broke we had to be able to fix it. That was interesting having to fix a GC in the field. They were old GCs as well, Agilent 5890s.

ABV by GC
There are a number of ways to measure ABV in beer, but we'll focus on gas chromatography for now (Method D). For this method, n-propanol is used as an internal standard and calibration standards of ethanol solutions within the range of the analyzed beer ABV are used. Beer does need to be decarbonated, and the method is relatively quick.
https://www.asbcnet.org/Methods/Methods/Beer-4.pdf
08/31/23
You'll need an ASBC membership to access

17.3 Current position

17.3.1 What are your current job responsibilities and tasks?

I do everything on the production side of the brewery. Cleaning and filling kegs, monitoring fermentations, recipe generation and all the brew day stuff. I also do quality control and sensory evaluation. Pretty much the whole thing. From a science background, what got me into the industry side of things was my boss in the oil field. He mentioned that he homebrewed and we bonded over that during my interview. So, beer got me a job in the oil field. We would talk about brewing and we would talk about homebrewing. He and his former business partner had this idea of starting a brewery and they would take the beers and analyze them on the GC for ABV content. When I was off-shift with nothing to do I would also look at what I could do with a

GC. I looked into the ASBC methods then and that is how I got started. Interesting note, his former partner was Matt Stinchfield who is one of the only level four cicerones in the US. He is the head of the Brewer's Association Safety Committee. That was cool to learn. I got laid off in the oil field in 2015. In Portland I started a contract QC lab for breweries. I bought a GC-MS, spectrophotometer, microscope, some gram staining equipment and supplies as well as a pH meter. I started to go through the methods. That was my first start in the industry. There are all these craft breweries, particularly in Portland, and so my thought was rather than making beer that I would work on farming myself out to breweries on a contract basis. If I could split my education and my time between 20–30 breweries then I could make it cost effective for the breweries and make some money as well. So, I did not have my girlfriend at the time and so I was bad at selling and ended up shuttering the business in 2018. I went back to manufacturing and now I'm in Idaho and had the chance to start this brewery. It was a twisty, windy road moving into the brewery business for me.

Cicerone
This program trains individuals on four different certification levels: Beer Server, Certified Cicerone, Advanced and Master. The Advanced Level is "a designation of distinctive expertise and tasting skill" while the Master level is the "ultimate test of beer expertise." Areas of the Master Level include serving beer, beer styles, flavor evaluation, ingredients and beer pairing.
https://www.cicerone.org/
08/31/23

17.3.2 What is your favorite and least favorite process to do and why?

My least favorite thing when I had that lab was the bake down and changing of columns. It was not fun because I always seem to get the GC ferrules wrong and so you end up with a leak and you need to let everything cool down again. That was very tedious and GC-MS was even more finicky and cleaning those was a real pain. My favorite thing to do was measuring the OG and more importantly measuring FG and actually hitting the numbers. Now that is cool. Taking that first sip and step is just great and my favorite part of the process.

Baking GC column
The process of baking gas chromatography columns assures that the columns are ready for component separation and that results will be consistent between samples. It also helps to remove potential contaminants from the column surface itself.
https://phenomenex.blog/2022/06/28/all-about-performing-gc-column-bake-outs/
08/31/23

17.3.3 What adjustments did you need to make from working in a lab setting to a brewery setting?

I think working in an oil field setting it wasn't a traditional lab setting. It is very blue collar and you are down to earth maybe and not stuffy. Coveralls and mud boots and not white coats. In a brewery it is similar and you are in a production atmosphere that is part of it and you are not ancillary to it. I am actually going to a hop processer to teach folks who are producing the hops more about the beer production process. Having a lab person that does not do homebrewing needs to be addressed – they need to care and know about it at least a little bit. We know somebody who focuses on training everybody, sales and otherwise on all the aspect of beer which I think is great. I also think having a lab person with that background in a brewery is very important. Also, the real-life application versus just theory.

17.3.4 Do you have any words of wisdom for students wishing to apply their chemistry (STEM) degree toward getting into the brewing industry?

I would say to just do it. We are kind of past the days of volunteering at a brewery. You can go in, apply at a warehouse and work as a cellarman cleaning kegs. For me up in Portland there was a lot brewed. Do not wait until you graduate to get involved. Get a job while you are at school. Even if it takes you longer you have that practical experience rather than just having a 4.0 GPA.

17.4 Experiences in the brewing industry

17.4.1 How would you describe your leadership style?

I am a direct leader and the collaborative leadership is my girlfriend.

17.4.2 What are some unique strengths that you think environmental science folks bring to the brewing industry?

I would say that broad knowledge and understanding. The difference is that brewing touches on a lot of STEM fields from the biology of yeast, to biochemical enzymatic processes to the chemistry of water. It is definitely more organic chemistry if you wanted one specific discipline. You still have logistics and even statistics is important. If you don't measure it you can't manage it. You need to be able to calibrate pH meters, look at a microscope slide, do gram staining and those are basics for a brewery from the science side. If you are going to work for a big lab a lot of learning how to run GC-MS or the Anton Paar and you'll be trained at a brewery. Having that understanding of how yeast strains will evolve and how to monitor and record it I think environmental science folks have that background for sure.

17.4.3 Have you had any negative experiences in the brewing industry that you would like to share?

Luckily I don't have too many. Charlie Papazian said at a Keynote once said that the brewing industry is 99% asshole free. There is still that 1% out there and fortunately most of them do not last too long. I think it's important to be collaborative rather than competitive. I know there is a tendency in academics to be insular and hush-hush about their research. You need to not do that in the brewing industry. In Portland there is competition, but I could still walk in and ask them for their recipe and they would probably give it to me. They'd talk about the grain bill, hop schedule and the like. In theory I could make that exact beer, but I just can't call it the same thing. The beers would definitely turn out differently. We would call these "clones" in the industry. Of course, nobody would market a clone of another brewery's beer though. There would also be that social stigma of just doing someone else's beer. The industry is very collaborative. We have plans for collaborations with local friends.

Charlie Papazian
I've actually heard him referred to as the father of homebrew/craft beer. His book *The Complete Joy of Homebrewing* is now in the fourth edition of printing – definitely worth getting if you're an avid brewer.
https://www.smithsonianmag.com/arts-culture/charlie-papazian-sparked-americas-craft-brew-revolution-180974877/
08/31/23

17.4.4 Do you have any tips or tricks for maintaining a positive work-life balance?

I would say to stand up for yourself and set boundaries. You have a first allegiance to yourself and your family. If the owner is asking you to come in on the weekend just say no. There are times when you do have to work long days, but that should come with a balance. "From those that can give a lot, much will be taken." I forgot where I heard that from. That is what I would say is don't be afraid to set your boundaries. Anyone with a STEM degree can get a job somewhere else.

17.4.5 Describe your toughest day in the brewing industry so far

I would say the first or second batch on the system for sure. It turned into a 14 h brew day. I'm trying to learn the system and nothing is going right at the same time I am trying to make a beer. You are constantly fighting pumps that are losing prime and you end up with a stuck mash. Anything that could go wrong did go wrong. My defense to that is you expect the worst and it doesn't happen and that is great. Part of adjusting is just experience and being able to turn the dials on the fly.

17.4.6 Anything else you'd like to share about brewing in Idaho or your area?

I would say that you need an RO system. Water will affect your beer more than anything else and you need to calibrate your pH meter.

17.5 Professional affiliations

17.5.1 What current professional organizations are you affiliated with? Why did you join them?

ASBC for sure. I was a member of the BA and I let that one lapse because we are not attending CBC this year – we don't have the bandwidth right now. I am also a member of MBAA, Northwest Chapter. I work with the surrounding homebrew clubs too. We are doing a wort share with Snake River Homebrew Club. I will brew the beer and then provide it to homebrewers and they can take it home and do what they want to it. They will judge it based on BJCP and I will judge and pick my winner. We will brew it on my system using the winner's method. We are a three barrel so it's not a big risk on our side.

Snake River Brewers
This club covers several areas in Idaho, including Boise, Nampa and the Treasure Valley. Their stated mission is "to further the interest and knowledge in the art of great fermentation."
https://snakeriverbrewers.org/wp/
09/07/23

17.5.2 What do you enjoy most about going to their events?

I like the collaboration and the creativity which is part of the alchemy. Asking people about experiences with yeast and hops and what they are playing with. Talking to people about the buffering capacity of the water since it's hard around here. When I had my testing company I liked to say that the brewing industry is a business done on a two drink minimum. It's very loose and social.

Alchemy
When people think of alchemy they almost always first think of turning lead into gold which is unfortunate as the article mentions. Alchemists focused on transforming matter into something else in addition to curing illnesses. As one of my professors said, "alchemists were simply doing chemistry when chemistry wasn't cool." This is a decent description of ancient alchemy, though some of the premises behind their version of chemistry have been found to be either inaccurate or impossible to do given our current knowledge of how elements behave.
https://www.livescience.com/39314-alchemy.html
08/31/23

17.5.3 How important is it to be associated with other professional organizations?

There are a bunch of different levels and I would say to get involved at the most local level as possible. That is going to be your immediate network and if you run out of filters you can borrow some from a bigger brewery and get it back to them. Widmer was even willing to run samples for you as part of their analytical process. Having that local network is very important particular for people just starting out.

17.6 Other/miscellaneous

17.6.1 Outside of brewing what other hobbies or interests do you have?

The big one was real estate investing. It got me into the brewing industry. I got into it in 2009 and long story short I was able to refinance some properties to buy the brewery and that cash flow allows the brewery to stand on its own, which is nice. It lets me brew what I want to instead of what I have to.

17.6.2 Is there anything else that you'd like to share about your personal or professional life that you think people might be interested in knowing?

I would say to never lose your curiosity and sense of wonder about brewing. It is a fascinating world and what goes on at the microscopic level should not take away from what is going on at the macro level. You brew but it's nice to go out in the taproom and see people enjoy what you made. Beer is said to be what helped civilization evolve. I would say to never let it become boring.

18 Morgan Hazard

BS in Biology, Minor in Chemistry TCU, 2013

Elysian Brewing, Quality Manager

TCU
https://www.tcu.edu/
08/08/23

Elysian Brewing
https://www.elysian
brewing.com/
08/08/23

18.1 College/degree pursuit

18.1.1 What inspired you to pursue a STEM degree?

Math was a natural strength of mine in grade school and science classes always kept my attention. One of my fondest memories of childhood was receiving a microscope as a birthday present. I would find bugs in the yard and examine them under the microscope – I had no idea how to use the microscope, but enjoyed playing scientist. As a child I had decided I wanted to be a veterinarian, then into my adolescent years that evolved to human medicine after an internship at a primary care practitioner's office. When it came time to start my first year at TCU, I decided to pursue a BS in biology because it would satisfy all my prerequisites for medical school through the degree path.

18.1.2 How long did it take you to graduate? Were there any stumbling blocks you'd like to share?

I graduated in 4 years. The transition from grade school to college was eye-opening. I struggled with focus for the first 2 years and had a hard time adjusting to living on my own. High school was not a challenge for me; I could procrastinate and still receive A's and B's, no problem. Once college coursework began, I lacked the self-discipline to study and prepare for exams ahead of time. Unfortunately, I received a D in biology my freshman year and ended up retaking it my junior year so I would have a chance at applying to med school. Although it was embarrassing for me to retake a course, it forced me to learn how to study and prepare ahead of time for exams/proj-

https://doi.org/10.1515/9783110798777-018

ects/presentations. This is a skill that has been highly translatable into my professional career.

18.1.3 Did you do any research as an undergraduate? If so, what was the project?

No I did not, but I was a TA in microbiology lab for a semester. When I took my first microbiology class it was as if the lights finally turned on in the house. The joy I found in learning about microbes led me to apply for the TA position in the micro lab and increased my desire to take more advanced microbiology classes. Through those classes and my TA position, I discovered the multitude of career paths available for somebody with a science degree. The food and beverage industry's many practical applications of microbiology were so fascinating to me. At that point, I wanted to learn more about the food and beverage industry but was not ready to change my medical career plans.

18.1.4 Do you have any words of wisdom for students wishing to pursue a chemistry (STEM) degree?

Yes, it is a very challenging degree path. Some people are naturally gifted at science and math, but others must work harder. Do not compare yourself to other students – every student learns in a different way. During my first 2 years, I did not take the time to meet my professors and participate in their office hours. Spend time building a relationship with your professors. They can help you succeed, as well as help you when you're struggling. That was a major lesson for me as a student.

18.1.5 How often do you use what you learned as an undergraduate in your current job?

Every single day. In my current position as quality manager of Elysian Brewing Company I oversee the microbiology and chemistry of beer. The reason why I landed in the brewing industry was my interest in microbes. Daily monitoring of beer chemical analysis allows us to react quickly to any potential issues with fermentation. The faster we can adjust the beer, the healthier the fermentation and yeast will be.

18.2 Movement to brewing field

18.2.1 What was your "entry beer" into enjoying craft beer? Why did you like it?

The brands I was introduced to may not be considered craft beer anymore, but I would say Shiner Bock and Blue Moon. Back then they were some of the only non-American light lager craft beers on the market. Additionally, while living in Fort Worth I was introduced to Rahr and Sons Brewing Company. It was a very popular local brand because the owner, Fritz Rahr, was a TCU alumni. His beer was on tap at all the restaurants and bars around campus and Rahr Blonde was my first true introduction. Craft beer was more appealing to me because it was more malty, sweet, full bodied, and more approachable than the plethora of American light lagers on the market.

Rahr's Blonde
Rahr's Blonde is a Helles Lager with an ABV of 5.0%. It is described as being "bright and crisp with a light body and mild finish with a subtle malt complexity."
https://rahrbrewing.com/
08/08/23

18.2.2 Did you get started with homebrewing?

No, I did not. I have dabbled in homebrewing Kombucha. I was only successful at growing the SCOBY (symbiotic culture of bacteria and yeast), but had trouble balancing the kombucha acidity.

Kombucha and SCOBY's
The SCOBY used to make Kombucha has several different names including "mother," "pellicle" and "pancake." It is added to freshly brewed tea to generate Kombucha. It may be convenient to think of a SCOBY as an analog to the sourdough starter used to make that style of bread.
https://www.youbrewkombucha.com/what-is-a-scoby
08/08/23

18.2.3 Did you ever brew a beer for an event (wedding)? How did it go?

No I did not.

18.2.4 Prior to getting into brewing what else did you do after graduating?

I took some time off after I graduated because I was debating whether medicine was the true career path I wanted to pursue. By chance I fell into an internship with Rahr and Sons Brewing Company in Fort Worth. I was hired to help the lab manager establish a microbiology program for the brewery. I did that for six months and then was hired on as their lab technician. I helped start the microbiology, sensory, and packaging quality program at Rahr.

18.3 Current position

18.3.1 What are your current job responsibilities and tasks?

I am the quality manager at Elysian Brewing Company in Seattle. My responsibilities include overseeing the day-to-day laboratory, maintaining brewing and packaging quality, and managing the food safety program of the facility.

18.3.2 What is your favorite and least favorite analysis to do and why?

My least favorite is IBU analysis. It is an easy procedure, but I am terrible at getting consistent results. I also dislike that I cannot multitask during the procedure. I love running analyses on the alcolyzer, but my most favorite of all still is microbiology analysis. Of the microbiological analyses, filter plating is my jam. There is something therapeutic about it. You can put your headphones on, get into a rhythm, and push multiple beers through a filter manifold.

IBU method (bitterness)
The ASBC method for measuring beer bitterness via IBUs involves a manual extraction using isooctane, 3 M hydrochloric acid, shaking apparatus a centrifuge and a spectrophotometer. Absorbance of the isolated extract is measured at 275 nm.
https://www.asbcnet.org/Methods/Methods/Beer-23.pdf
08/08/23 (requires ASBC Membership)

18.3.3 What classes that you took apply to some of the things that you have done in the brewing industry? Why/how?

Hands down Intro to Microbiology and in particular the lab portion of that course. I learned how to prepare agar plates, streak plates, staining procedures, and even how to make yogurt. We do not make yogurt in a brewery, but it is a method of growing a culture. Every brewery I've worked for has a propagation procedure where we grow yeast or bacterial cultures for different styles of beer. In the lab we prepare agar, aseptically sample, and plate beer throughout the process. We utilize a variety of differential and selective medias to observe microbial growth trends from brewing to packaging.

18.3.4 What adjustments did you need to make from working in a lab setting to a brewery setting?

There are limited resources in a brewing lab and you have to adapt to the environment that is provided. Universities have advanced technology and higher quality equipment. Most small breweries don't have money to acquire high quality equipment or to build a dedicated laboratory. At my first brewery job, our lab was in a supply closet in the corner of the brewery without air conditioning. Two of us barely fit in a room with a fold out table, sink, desk computer, a mini fridge, and an incubator. We made the best of the situation. Eventually, we added a wall A/C unit for temperature control. Instead of a laminar flow hood, we used a butane torch to maintain a sterile plating environment. We used a pressure cooker to autoclave our micro equipment. The lab I work in now has a true laminar flow hood, autoclave, as well as other nice equipment, but I know how to adapt if any of that equipment bites the dust.

18.3.5 What do you enjoy most about your current position?

I love my quality team and I love working with all the different departments – brewing, packaging, maintenance, safety, sales, and marketing. Quality works with the brewing team to give them in-process beer chemical analysis so that they can make critical decisions to stay on schedule. We are very hands on in the packaging department. Not only are my team and I down on the floor working with the packaging team to make sure the beer is meeting our packaging quality standards, but we are also making sure the finished product is within specification for shipping. I work with maintenance to fix equipment problems and establish brewing/packaging parameters. Safety is very important to our facility, so we are constantly working to make the production environment a safer place for our employees. I work with sales and marketing to resolve consumer complaint issues, as well as help make decisions on packaging design for new products. I have my hands in just about everything and I love it.

18.3.6 Do you have any words of wisdom for students wishing to apply their chemistry (STEM) degree toward getting into the brewing industry?

If you're looking for a career path that pays a lot of money, you're in the wrong field. To be happy and successful in the brewing industry, you must be passionate about beer. There are days with long hours working manual labor to meet a deadline, but those are some of the most rewarding days. If you're a hard worker, a self-starter, passionate about beer and looking for an easy-going culture – the brewing industry is for you!

18.3.7 What would you like to share from your experiences in quality management?

I have learned not to over-react to problems. When I get worked up, it clouds my judgement and I end up regretting my decisions. If I find myself at a dead end in a problem, I utilize my resources and reach out to my peers in the brewing industry. I'm very lucky in my position through Anheuser-Busch we have an amazing network of subject matter experts. Even at breweries outside the AB network, most everyone is so collaborative and willing to lend a hand.

18.4 Experiences in the brewing industry

18.4.1 How would you describe your leadership style?

Early on in my career I was a classic micro-manager. I would constantly watch over and correct the work of my employees. Not only did it take a toll on my mental health, but it was a waste of my valuable time. Over the years my leadership style has evolved into utilizing the strengths of my team members, giving them the opportunity to fail, but helping them learn from those mistakes. It has done wonders for my mental health, and it allows my team members to succeed and grow professionally.

Micromanagement
The majority of people working will likely tell you that they have been micromanaged at some point in their working career. While the descriptions vary, one common theme is the feeling that your boss does not trust you with your assigned tasks or making decisions without directly involving them. "Keep me in the loop" is a common phrase used by micromanagers. From my own experience, the best way to deal with micromanagement is to open up a dialog with your boss and attempt to develop a mutual understanding and respect for each other. For all you know their supervisor may actually be the micromanager, and they are just trying to meet their expectations.

18.4.2 What are some unique strengths that you think women bring to the brewing industry?

That is a good question. I think women are naturally more empathetic and some have a strength in building relationships. In my position, building relationships with other departments at the brewery is key to growing quality ownership of individual employees. When you build-up the trust of other departments, they will bring quality issues to your attention and sometimes bring a solution along with them. I have always encouraged my team to build genuine relationships with their peers.

18.4.3 Have you had any negative experiences in the brewing industry that you would like to share?

Many times. I have had to advocate hard for changes I believed in and my peers were not always professional. I am naturally a sensitive person, and there is nothing wrong with that, but I have developed some thick skin. Even though I had some negative interactions it made me a stronger person and a better manager.

18.4.4 Do you have any advice that you would give to your younger self before getting into the brewing industry?

Don't be afraid to be wrong. Don't be afraid to speak up.

18.4.5 Do you have any tips or tricks for maintaining a positive work-life balance?

When I go home, I do not look at my email or open my laptop. Those two things have been key to a positive work-life balance for me. My time at home is for my family, exercise, and relaxing. As a manager, I am on call 24/7 though. I will respond to a call or a text when there is an emergency at the brewery.

18.4.6 Describe your toughest day in the brewing industry so far

There have been many tough days and I can no longer count them on two hands. One of the earliest "tough days" I had was when I was working as a laboratory technician. We were transferring all our bourbon barrel aged beer for our winter seasonal, and this was the first year we were performing micro testing on a blended brite beer tank. I was new to the brewery and had just established the micro program. Since there was no history of micro data, management did not have much trust or knowledge in micro results. Unfortunately, the results came back in days with a *Pediococcus* strain contamination after we packaged and shipped out the beer. I was so nervous and scared to share the results with the production manager the day I discovered the results. We tasted our retained samples and the beer had soured within days of packaging. We ended up having to withdraw the entire batch from the market and it cost the company tens of thousands of dollars, if not more.

18.5 Professional affiliations

18.5.1 What current professional organizations are you affiliated with? Why did you join them?

I am affiliated with ASBC. I joined because my first manager was a member and had amazing experiences attending the conferences. The industry seems big, but it is a small world. Everyone seems to know each other, and most are willing to help solve a problem. Not only are there conferences to attend for networking, but they are the best resource for all new and established scientific methods for the brewing industry. I am not a member of MBAA, but I would also recommend joining if you're interested in the brewing technology side of the industry.

18.5.2 What do you enjoy most about going to their events?

Learning about the up-and-coming research is interesting, but what's been most applicable are the workshops led by brewing industry leaders. I have taken more practical knowledge home with me from those presentations than any technical talks. One of the best parts is drinking beer with people doing the same thing you are at another brewery and gaining new perspectives from them.

18.5.3 How important is it to be associated with other professional organizations?

I think it's very important to gain a community for both networking and problem solving.

18.6 Other/miscellaneous

18.6.1 Outside of brewing what other hobbies or interests do you have?

I love everything food related. I love cooking and testing new recipes from my favorite food bloggers. I am a sucker for all cooking competition shows, especially anything Gordon Ramsey creates. I also love researching and trying new restaurants in my area and anywhere that we are traveling. This year I am experimenting with growing some of my favorite herbs and veggies in my backyard garden.

18.6.2 Is there anything else that you'd like to share about your personal or professional life that you think people might be interested in knowing?

I am married and have a daughter named Porter. She is named after my great grandfather and coincidentally a beer style 😊. I grew up in Texas and am now living in Washington. I love Texas and it is a huge part of my life. It's where I grew up, went to college, visit my family frequently, and started my career in the brewing industry. In between my residence in Texas and Washington, my husband and I moved to Arizona for 5 years so he could complete a graduate degree. I was able to continue my career in brewing at Four Peaks Brewing Company as their quality manager. During Covid we moved to Washington state, and I am now working as the quality manager at Elysian Brewing Company. Living in these three states have shaped who I am, brought different perspectives to my life and career, but I will always be Texas girl at heart. Gosh I miss my family and some good BBQ!

Types of BBQ

Okay, so we will probably never agree on the exact number of BBQ types that exist in the United States or, for that matter, which is best. The cited article mentions over 15 styles, 3 of which are from Texas! I'm getting hungry for some just typing this.
https://www.tastingtable.com/695400/styles-american-bbq-barbecue/
08/08/23

---

19 Tyler Mappe

– *Baylor University, 2015, BS in Biology*

– *Canarchy, Quality Director*

Baylor University
https://www.baylor.edu/
08/31/23

Canarchy
https://www.canarchy.beer/
08/31/23

19.1 College/degree pursuit

19.1.1 Where did you get your degree from and what inspired you to pursue it?

Baylor University. I actually got a degree in biology, but technically was on a prepharmacy track. During the last 2 years I became more interested in craft beer and styles and learned that breweries had labs and that seemed pretty enticing for me. The universe played into it a bit, so I decided to try a different path after that.

Prepharmacy track
Some of my biochemistry advisees have pursued pharmacy school. My best pieces of advice are to look at the school(s) you're interested in specifically for entrance requirements and consider becoming a pharmacy technician first.
https://blog.collegevine.com/pre-pharmacy-requirements
08/31/23

19.1.2 What inspired you to pursue a chemistry (STEM) degree?

That's interesting. I've always had an analytical nature. My personality has been that way. I like to dissect things and learn how things work and kind of grew up that way. I thought in my youth that I wanted to go into the medical field. I'm going to go into this lucrative field and help people. Then reality takes you on its own plans and so I feel like I've been satisfied in my field and that I would have made as much money as going into the medical field but honestly it kind of took me that way. I started to learn about the food and beverage industry and the application science. I just kind of fell in

https://doi.org/10.1515/9783110798777-019

love with that and just started working instead of chasing something (professional school) that might not have worked out in the long term.

19.1.3 How long did it take you to graduate? Were there any stumbling blocks you'd like to share?

Eight and half semesters. Yeah, those upper class years where you get into the advanced courses. It starts to get hard at that point if you're not a focused student you'll fall behind especially at a competitive school like Baylor. One that I was going to where the science program was pretty focused and interested in helping people get into medical and professional schools. I mean I thought that was a bonus for me getting into the brewing world. I felt like I had advanced microbiological knowledge and understood the chemistry and physiological processes that people with degrees in other areas might not have. There is that and ultimately I felt pretty fulfilled in doing the job there. Once the concepts start getting difficult to follow you can lose focus and get distracted. I mean I guess if you're asking the question for incoming students, it wasn't until halfway through my first year where trying to build a brewing lab which is where it all started to click. Like this is what I learned this for and it's applicable. You do end up appreciating those courses like organic chemistry, but you may not see it when you're going through it. You will get to the point where you're trying to calibrate a UV–Vis measurement method and you realize that your education really helped you.

19.1.4 Did you do any research as an undergraduate? If so, what was the project?

I did not have the opportunity to do research. Looking back on it I wish that I would have. There were some opportunities. I could have done a small project and just didn't take advantage of that. In hindsight I wish that I would have.

19.1.5 Do you have any words of wisdom for students wishing to pursue a chemistry (STEM) degree?

I guess really you have to be passionate about it. There's a lot of people who are interested in it but they don't have a passion for it. I was fortunate because some of the programs look to weed out students, but I benefited from being hooked on it. If you find yourself going to every single office hour and spending time with the professors and struggling and not connecting with the professor then it may not be for you. Just go to the office hours – stuff will be really hard. Work through it, send emails to your professors. Don't be shy. You can also find other people to rely on. Don't isolate your-self and you can succeed.

19.1.6 How often do you use what you learned as an undergraduate in your current or former jobs?

Every single day. Right now I'm going through a microbiological investigation. The drinks I work with are fruit juice based. We're dealing with the same contamination concerns that you would in beer processing. It's just a different species of organisms that you're dealing with. Beer producers look for *Lactobacillus, Pediococcus,* etc. while fruit juice-based beverage producers are worried about spore-forming, heat-tolerant *Bacillus* species (*Alicyclobacillus*) which is what I'm finding out. It's basically like all of my upper level micro classes. I use them every day. Bacterial differentiation, identification with selective medias, etc. In the beer world, UV–Vis was king. The methods are cheap and very easy to run once you get them set up. We also had GC-MS at Canarchy with the resident chemist in Longmont who would do headspace analy-sis, Robin Mead, he's the most tenured guy there. He is super knowledgeable.

Alicyclobacillus
This gram-positive, rod-shaped bacteria is a common spoiler of fruit juices and can form spores. Off-flavors associated with these are typically of the phenol variety.
https://www.sciencedirect.com/topics/agricultural-and-biological-sciences/
alicyclobacillus
08/31/23

Selective media
It seemed a shame to just focus on media, so here is a full list of supplies from White Labs – navigate to Prepared or Dry Media in the search area. Selective media does what it suggests and allows the lab technician to rule out or not eliminate certain types of microbes (e.g., wild yeast, *Lactobacillus* or *Pediococcus*).
https://www.whitelabs.com/product-search?page=2
08/31/23

19.2 Movement to brewing field

19.2.1 What was your "entry beer" into enjoying craft beer? Why did you like it?

Belgian Trippel was the entry beer. Like all the ones out there. I was fascinated with the Belgian style. You just get interested in other styles from then on. As a new craft beer drinker I really liked it. Really sweet, high alcohol, fruity unique character. More approachable than heavy bitterness, very dry beers that tend to kick non-experienced beer drinkers to the curb.

Belgian Trippel
This beer belongs to #26, the Monastic Ale category. As it name implies, it does pack a punch with ABV levels ranging between 7.5% and 9.5%. St. Bernardus and Chimay are two very good commercial examples of this style. Comments regarding this style are that it is "High in alcohol but does not taste strongly of alcohol. The best examples are sneaky, not obvious. High carbonation and attenuation helps bring out the many flavors and to increase the perception of a dry finish."
https://www.bjcp.org/style/2021/26/
08/31/23

19.2.2 Did you get started with homebrewing?

No. I actually have never done it in my whole life. Never was really interested in doing it. Like to see a small batch be run on a small system and to see people struggle with why it didn't turn out the way they like. I'm a professional brewer and so I decided to just stick with that.

19.2.3 Did you ever judge beers?

Only like at the local festivals. Never really judged at a higher level.

19.2.4 Prior to getting into brewing what else did you do after graduating?

I was a pharmacy tech and worked part time the last years that I was in school. You do get a sense of working with the public in that sort of job. Whereas talking to people and making sure they're satisfied with what they are receiving and that sort of thing. I thought that was a very valuable experience. You're going into a place where you are interested in the consumer experience. Making sure people are happy with what they're getting. Even as you're handing it to them across the counter. I didn't do much bartending. You do want to make sure that it is high quality that you are giving to them.

19.2.5 Would you recommend brewing to other chemists out there and why?

Honestly, yes if you're really interested in craft beer and you've fallen in love with it and you're willing to sweat constantly and work in an environment where you are in

industrial manufacturing. It's not like you'll be in a desk job. It's going to be rewarding and brings value to their life. You will work hard and sweat every day. It will be
more physical than you may have experienced while working in a lab. If you're ex-
pecting to go in and wear a lab coat and goggles that is probably not going to be the
case. You'll be wearing steel-toed boots. You'll probably still be wearing eye protection. But you'll be taking a sample with sweat dripping down your forehead, holding a
torch in one hand and a sterile tube in the other. The reality is it is hard work and
people make such an incredible drink at the end. And you just have to be fascinated
with the microbiological processes. So, yeah the passion has to be there for real.

19.2.6 What similarities and differences would you draw from the beer world and the seltzer world?

We function primarily as a mixing facility. We primarily blend distilled spirits, mostly
tequila with fruit juice, blend it in a tank, pasteurize it and place it in a can. It is very
simple compared to beer brewing. There are still concerns about product longevity
and that sort of thing. I was just talking about microbiology issues. They are similar in
that you need to have a good understanding of how all the machines work and how
to send fluid through the pasteurizer and through the filling machine and then out
the door. So that it can sit on a shelf for a year and be fine. If you're working in a
distillery because there is still a lot of cross over because you're also starting with fermentation. You are measuring yeast health. We used fluorescence microscopy to measure yeast health at the brewery. That was our other big tool that we used next to
UV–Vis spectroscopy.

Seltzer production
Seltzers have become a very popular choice among some beer drinkers as they are
lower in alcohol and calories. I keep waiting for somebody to come out with a "double
seltzer" or something similar. These are very easy to make, requiring only water, sugar,
yeast and some type of flavoring agent.
https://firstkey.com/hard-seltzer-production-methods/
08/31/23

19.3 Current position

19.3.1 How did you end up in the brewing industry

Basically after I graduated I emailed practically every brewery in Texas and found
one that was willing to interview me. I ended up in Dallas at Deep Ellum because they
needed someone to help kickoff the Quality Program for them. Basically, that was the

first half of my training. The more immersive part of it first happened at Baylor. During my stay at Deep Ellum they were acquired by Canarchy as in Oskar Blues. There's a ton of really experienced chemists and brewers there that really helped me out. I did have ASBC which was a crazy resource for me as far as how much information I could get. I was new and green to the whole field. I understood how to build a lab at that point, but I didn't understand how to be targeted in beer quality control and ASBC really helped me out. These are some things to target for. There was a graph that described what equipment you should buy for your beer lab and had a plot or a graph for what you should use. You could just look at that and you could be like I'm trying to find what equipment and these are the things I should be testing for based on size of my brewery. That was very helpful, but I mean after learning from the experienced brewers and the other lab managers in the company you start to realize how much your upper level micro classes are super applicable to the brewing world. After that I was hooked.

19.3.2 What are your current job responsibilities and tasks?

At Deep Ellum I was in a middle manager position. I managed the lab and then I also managed the lab technicians. I had two at one point and then we streamlined the positions using one technician. It was mostly gathering data and giving it back to the brewers and the operators and then saying here is what I think went wrong and it was recommending an action based on what the lab in your facility has gathered. The difference now is that I basically direct the whole Quality Program and if I want to put a full stop on a process, I can. If I want to change a process specification because I think it will bring long term quality and stability to the product after I read a paper, I can just do that. I don't have to go to ask anybody or that sort of thing. So I enact the quality policies and specifications instead of just feeding the data back to the brewers in a sense.

19.3.3 What classes that you took apply to some of the things that you have done in the brewing industry? Why/how?

All of my STEM classes. Biology 1 and 2. Chemistry 1 and 2. Physics 1 and 2. Freshman and sophomore level. Organic 1 and 2. Your upper level chemistry and physics courses. I really enjoyed the upper level classes though. Those are really applicable for me in brewing in terms of evaluating yeast health, understanding how the yeast behave in the beer. Potential bacteria spoilers and how they might interact with the yeast are also important. How to prevent them from being there. Different spoilers that you are looking for. It really helps you get that different level of understanding as to how those microbiological processes work. Tying into the off flavors in your beer and where they come from biochemically are very helpful, right. That is where your organic chemistry, biochemistry and physiology of the cell comes in. Even though my education focused on medicine it was very applicable to understanding yeast function and their metabolism and what they can do. Also, how to understand why you have higher VDK levels – how do we modify our biological process rather than adding enzymes.

Bacterial spoilers in beer
A number of microbes can spoil a brewer's day including Lactobacillus (think lactic acid) and Brettanomyces ("horsey off-flavor"). A final type is Enterobacteriaceae which actually spoils wort as opposed to beer. Proper cleaning and sanitation are the two best tools to use to avoid microbial contamination.
https://brewingscience.com/beer-spoilage-organisms/
08/31/23

19.3.4 What adjustments did you need to make from working in an undergraduate lab setting to a brewery or industrial setting?

It was hard at first for sure. I'm moving from an academic, planned experiment where the outcome is predictable to where you're analyzing something and trying to get some reproducible results that are consistent. It goes back to what I was saying earlier. You can't be afraid of sweating or getting dirty. You have to put some elbow grease into it because that is when you're really going to find and shine in your abilities. It probably took me 6 months from the time that I began that job to getting the lab up and running to before I got a consistent VDK value or number. It takes some time and it's like a thousand tiny cuts. Every brewery is going to be like that with the compulsiveness and dealing with things like that.

Beer off-flavors
Beer can have several off-flavors that can be attributed to malts, microbial contamination, issues with fermentation or water source. Dimethyl sulfide, for example, comes from the grain source and tastes like creamed corn. Wet cardboard is associated with beer oxidation, while skunkiness is due to exposure of fermenting or finished beer to UV light.
https://www.cicerone.org/sites/default/files/resources/off_flavor.pdf
08/31/23

19.3.5 What do you enjoy most about your current position?

What drove me to this position was two things. The energy and momentum that I saw in the brewery world and what I saw in the mixed drink industry. I mean I really love how many scientists there are in the brewing industry. So many excellent people and intelligent people are doing beer and beer flavor research. I moved into the other industry because the pay was a little better in the other sector at least in my area. I do wish that some areas of the brewing industry would pay their operators more to help drive talent in the industry.

19.3.6 Do you have any suggestions for students wishing to apply their chemistry (STEM) degree toward getting into the brewing industry?

If you want to have a job lined up then it's fairly easy to go around and talk to your favorite breweries. Even your large ones will talk to you. Even some bartenders or people in the tap room can get you the contact information. People will take you seriously if you are passionate about getting into beer. In the brewing industry there is a tremendous camaraderie among brewers. Even across the big brewers to the small brewers there really are no secrets. Everybody wants to help you and work with you. Just keep reaching out and making friends with your local craft brewer. They probably want someone to like shovel out their mash tun. I mean I've been the guy that just comes in and sits at the end of the line and presses the tags on the six packs. You can just go in and say "I'm just finishing my degree and I'm interested in learning what opportunities you have around here and that starts the conversation, right?"

19.4 Experiences in the brewing industry (not covered for this interview)

19.5 Professional affiliations

19.5.1 What current professional organizations are you affiliated with? Why did you join them?

Only ASBC right now. I've been in this position for about 6 months but am looking for others to join. When I joined them I was looking for a resource to start the brewing lab in Dallas and basically there was nothing else I could find out there. There were research papers and so you had to wade through them to get what you needed, right. ASBC provided structural thinking and a benefit where you can be a brewing chemist just starting out. You know like this is just what I need starting out with a smaller brewery going into a larger brewery. There are resources along the way to help with that. It's just like super friendly for me.

19.5.2 What do you enjoy most about going to their events?

I was at the San Diego Brewing Summit and was also at the New Orleans meeting. I kind of miss it. Honestly, the meetings are very enjoyable because you really engage your intellect. The amount of research that is there is super engaging. You expect to get there and it's all boring but the amount of applicable research that is being done that you can like take back to the brewery and use almost immediately is pretty insane. I'll just say that. The second part is networking. You want to have friends in the academic and the professional community. It is always going to come in handy for you. Even if you're a guy and you are wondering why you have floaties in your dry hopped beer and you can contact the guy at another brewery and they can help you with that. That was actually something that I needed to understand and it was extremely valuable for me.

19.5.3 How important is it to be associated with other professional organizations?

It can be helpful for students, but as soon as you get that position as say QC manager it's even as important. Even if you're a cellar head or head brewer you should get that subscription. There is tons of information out there. The cost of it is very cheap compared to everything else that you learn.

19.6 Other/miscellaneous

19.6.1 Outside of brewing what other hobbies or interests do you have?

I am a very casual mountain biker. Don't take it hard like I used to. Another hobby is cooking. Beer is the gateway to the foodie. You become a foodie after you get into craft beer. So yeah cooking is fascinating.

19.6.2 Is there anything else that you'd like to share about your personal or professional life that you think people might be interested in knowing?

I guess the tag line "In my life when I've been making personal career decisions go with your gut." It has been helping me so far.

20 Monica Mondragon

BS in Anthropology with a Minor in Biology, 2005, UNM in Albuquerque, NM

MS in Biological Anthropology – Focus on Biological and Forensic Anthropology, UNM in Albuquerque, NM, 2008

Steel Bender Brewyard in Los Ranchos De Albuquerque, NM, Operations Manager

UNM
https://www.unm.edu/
08/10/23

Steel Bender Brewery
https://steelbenderbrewyard.com/
08/10/23

20.1 College/degree pursuit

20.1.1 What inspired you to pursue a STEM degree?

I have always been interested in the scientific track even as a child. I was very inquisitive. I like the deductive reasoning and the scientific method. It really resonates with me. I was going to go into music theory because it is math based, but then I went the biology route. To expand on the biological anthropology track it started focusing on human evolution. Then it eventually developed now into looking at interactive human choices and the science of that. It has moved less from sort of historical to more like life histories and patterns through physical means. How is everyone changing because we are moving around less? How does that change your bone structure due to your activity patterns?

20.1.2 How long did it take you to graduate? Were there any stumbling blocks you'd like to share?

It took about 5 years. I did have a little stumbling block – I was planning on going toward the full biology track. I learned that organic chemistry is not my friend. So then I found out about the biological anthropology route. I have always been kind of interested in anthropology so I could blend a social science with an actual science. It really appealed to me.

https://doi.org/10.1515/9783110798777-020

20.1.3 Did you do any research as an undergraduate? If so, what was the project?

I actually did. I did a senior project looking at bone density of different professions. So, there was a really good paper. They had been looking at this in historical samples. Looking at the muscle attachment points on their bones in relation to their activity levels. We had a sample size of modern volunteers who agreed to be part of the study. We were able to look at bone density and then looked visually through slides. We wanted to see if we could tell if people who had more active professions like furniture making had denser bones compared to those who sat at a desk all day like a teacher. We saw some differences but the sample size was not big enough to see a difference. We did see some interesting patterns though.

Bone density measurements
Bone density measurements are very important to older individuals to detect osteoporosis before a bone is broken. It determines how much calcium, and other minerals, are present in bones and is measured using dual energy X-ray absorptiometry (DXA). An advantage of this method is that it is noninvasive and is relatively quick. Additionally, the amount of radiation used is significantly less than what is used for standard X-rays.
https://www.bonehealthandosteoporosis.org/patients/diagnosis-information/bone-den sity-examtesting/
08/10/23

20.1.4 Do you have any words of wisdom for students wishing to pursue a chemistry (STEM) degree?

Yes. There is always going to be a class where you may not quite get it. You may struggle with it, but stick with it and take the time and find people who can help tutor you. The can help you get you a better understanding and help you retain that information better.

20.1.5 How often do you use what you learned as an undergraduate in your current job?

I try to take that scientific method and apply it to the production floor. Especially when we're having fermentation issues, carbonation issues. I don't want to assume that I know what is going wrong so I go through all the little pieces to figure out what might actually be causing the issue we are looking at.

20.2 Movement to brewing field

20.2.1 What was your "entry beer" into enjoying craft beer? Why did you like it?

Probably before I was legally supposed to drink it. Early on in college, I always preferred the more flavorful characteristics and the different varieties and styles of craft beer. In 2011, I started a craft beer enthusiast club. We were a local, grassroots club and the craft beer breweries were just getting started in Albuquerque. We wanted them to stick around and so our philosophy was to get a whole bunch of us together when they are not busy and try to give them some business to help keep them open. My first craft beer was Pete's Wicked Ale. At the time the parties only had the commercial lagers that were really light and somebody handed me one and it was hoppy and malty. I was like "what is this, I have never had a beer that tasted like this before." That is what started it all.

Pete's Wicked Ale

This beer was produced by Pete's Brewing Company which sold out to The Gambrinus Company in 1998. Anybody who was drinking craft beer in the 1980s or early 1990s will likely remember this beer. Pete's Wicked Ale was a Brown ale with an ABV of 5.5 (Beer Advocate) that ceased being produced in 2011.

https://www.beervanablog.com/beervana/2011/05/fast-rise-and-slow-death-of-petes.html

08/10/23

20.2.2 Did you get started with homebrewing?

Yes, actually. My current husband when we were dating he had started homebrewing and so we got started in it together. He is now in the beer industry as well doing sales so it is all in the family.

20.2.3 What was your first recipe?

I do not remember the actual first one. We really enjoyed doing stouts and the more maltier beers. You know stouts, ambers and nut browns. At that time we did not have the equipment needed to make lagers to even attempt. And IPAs were not around yet.

20.2.4 Did you ever enter a recipe into a brewing contest?

No I did not.

20.2.5 What was your biggest disaster?

We tried to brew an Imperial Stout. We made a recipe and it turned out really good. We tried to repeat the recipe and it ended up as a stuck fermentation. We ended up with a real sugary beer. That one never fully attenuated and that one really hurt. You know the grain bill for an Imperial is not cheap.

20.2.6 Were you part of a homebrew club? Which one? How long a member?

That I did not.

20.2.7 What was your biggest batch as a homebrewer?

We were doing two separate batches a weekend so probably 10 gallons. We were doing 10 gallons a week for a couple of years there.

20.2.8 Did you ever brew a beer for an event (wedding)? How did it go?

Nope.

20.2.9 Prior to getting into brewing what else did you do after graduating?

It was a slow transition. I started off in my field doing contract archeology to start. From there I moved into the forensic side of anthropology. I worked for the medical examiner's office for 6 years working in the autopsy suite and in the lab. Most of the work I did in anthropology was lab work and using lab protocols and lab procedures. So when I left the medical examiner's office in 2014 that is when I got into the brewing industry at Santa Fe brewing company as their QA/QC lab manager.

20.3 Current position

20.3.1 What would you like to share about your experiences in QA/QC at Santa Fe?

If you are coming from a scientific background it is very important to understand that the way that these types of labs work for this type of manufacturing is very different from the typical scientific lab work. That can sometimes be a shock for people if they just got started in their career. You know in the scientific lab protocols are pretty stringent with working in a chemical or biology lab. If you are working in QA/QC on a smaller production level as most of the smaller breweries are there are quite a few things you have to forgive. You need to understand that the processes are not the same because it is a little bit different since you're working on a beverage.

20.3.2 What are your current job responsibilities and tasks?

I got promoted to operations manager and right now that entails everything that happens on the production scale which includes everything from the brewing floor to the cellaring and packaging schedule. Also scheduling beer releases. I work very closely with the sales team to make sure that their sales forecasts get executed. I also help in developing recipes, writing standard operating procedures (SOPs), inventory, management and purchasing.

20.3.3 Of those responsibilities you mentioned which were the most challenging in making the transition from QA/QC to operations?

I think it had to do with understanding the types of reports and reporting what is needed on an upper manager level in running a brewery. When you are working the production floor you don't know anything about Alcohol and Tobacco Tax and Trade Bureau (TTB) taxes so that and others things are new. I'm still trying to wrap my head around it a little bit.

TTB
The TTB regulates the brewing industry at the Federal Level. They are part of the US Department of the Treasury and oversee beer, distilled spirits, wine, sake and kombucha. This includes beer labeling, importing/exporting, taxes and operations-related items. Individuals thinking about starting a brewery need to consider the TTB application process prior to setting an opening date.
https://www.ttb.gov/beer
08/10/23

20.3.4 What classes that you took apply to some of the things that you have done in the brewing industry? Why/how?

Basic Cell Biology 101. We learned everything about Cell Biology the primary example was the yeast cell. It's also used in brewing so things like the Krebs Cycle are pretty much what we are doing in the brewery.

20.3.5 What adjustments did you need to make from working in a lab setting to a brewery setting?

In a lab setting it is all about data collection and all data can be relevant compared to what you are trying to achieve. In a brewery setting it is more of what is most actionable. Most brewers don't want just data, they need specific data and data they can act on in real time to help improve what they are doing. There are some things in a lab background that you may want to do in order to get this information, but in a brewing lab they are not interested in some of those things because they cannot control them. So in brewing they want to focus on what they can act on so you have to go with the flow like that. It's especially true when you are working with a brewmaster who went to brewing school and has some scientific knowledge. Others do not have that background so it kind of depends on the brewery that you are at and who your brewmaster is.

20.3.6 What do you enjoy most about your current position?

What I like about operations is that you have to be part of everything in order to make it work. I like being part of that upper level management that helps to make decisions and make things more efficient and helping the brewery improve efficiencies.

20.3.7 What would you say you enjoyed the most about the QA/QC position?

I felt like I had a lot of pride in helping to make the product better. Putting the processes in place to make a good quality product and being able to monitor it on a daily basis. Troubleshooting something was one of my more favorite things to do in the brewing process. Yeast management was probably my most favorite thing in that role.

20.3.8 Do you have any words of wisdom for students wishing to apply their chemistry (STEM) degree toward getting into the brewing industry?

I would say do not be afraid of taking some biology courses. Many of the people I have run into have taken chemistry and while there is a lot of chemistry in brewing, I think sometimes they forget that it is a biological process they are working with. *Saccharomyces* is a fungi that is a living, biological thing that we are manipulating to get what they want out of it. So I think it's important to take that Cell Biology 101 class and learn about cell membranes and how they uptake chemicals and all of that important stuff.

20.4 Experiences in the brewing industry

20.4.1 How would you describe your leadership style?

In my working career I have kind of been keeping tabs on what I liked and did not like about managers. I tried to not do those things I did not like. I prefer to do one on ones with my employees. I like to know where their head is at and give them feedback about what is going on with their jobs. What I did not like about those annual evaluations is that they hit you with a bunch of stuff that you could have been working on to improve throughout the year. On the other hand, if you had been sat down and told that you need to improve things in this area it could have been acted upon. I also like a lot of teamwork. Brewing is a team situation and so we all have to work together. We have to work efficiently and I try to give them as much information as possible so that they can do their jobs properly. I tend to be SOP heavy not for training but for reference purposes. That is because we all forget something occasionally and if you are here on a Saturday by yourself and you need those to help you figure out what it is that you need to do.

SOP
SOPs are the industrial analog to the laboratory protocol except with much more detail. The FDA Group provides a model for effective SOPs which include the following items: Procedure title, Purpose, Scope, References/Related Documents, Definitions, Roles and Responsibilities, Procedure, Appendices, Revision History and Signatures.
https://www.thefdagroup.com/blog/a-basic-guide-to-writing-effective-standard-operat
ing-procedures-sops
08/10/23

20.4.2 What are some unique strengths that you think women bring to the brewing industry?

That's a great question. I think particularly with the ones that I have worked with they tend to be more collaborative. Every single one that I have worked with and been acquainted with are all very collaborative. They like to ask a lot of questions which sometimes can be annoying for certain personality types, but I do see some more collaboration if there tends to be females on the team.

20.4.3 Have you had any negative experiences in the brewing industry that you would like to share?

I would have to say that it is not really an industry to get into if you plan on making a lot of money. It is a trade, a specialized trade but it is a niche industry. You see these bubbles burst and expand. Beer is one of the few alcoholic beverages where it tends to happen. So you have to be prepared for dry and slow times. The hours can be long, especially in the summer. I am pretty much on call a lot of the time. Even when I am off because problems arise and you have to jump on them any time that you can. Basically, if you want to succeed in the industry you have to be part of all of that stuff. I know some people who thought that they wanted to work in the industry and they realize what their pay is going to be. We do have a great time but it is very physical work. When I was at Santa Fe they went up to 40,000 barrels the year I left. So you are on the brewery floor all day pulling samples and doing tests. Yeah it is a lot more physical work than people think.

20.4.4 Do you have any advice that you would give to your younger self before getting into the brewing industry?

I don't think so honestly. I generally try not to look back that far. So I can't change any of it so it is what it is. I tend to focus more on going forward and what I am going to be doing in 3–5 years. I actually have not thought about that much. If I had to give myself some advice it would be to lighten up a little bit. I went into a brewery lab coming from working 10 years in a science lab and I had to keep reminding myself that this is just beer. There are little things in a science lab that we would have to address immediately so, yeah, I would say to just lighten up a little.

20.4.5 Describe your toughest day in the brewing industry so far

I would have to say when the seamer failed on our filler and we had to go back through pallets and pallets of beer to figure out when it happened. We then had to dump whatever it was that we found that had an issue. It was quite a number of pallets to say the least. It was probably one of the hardest days. So I tell everybody now to get your seamers checked. Once it fails you are down for at least a week.

20.4.6 What would you say are kind of the current style trends in the industry?

I am seeing a lot of N/A (no alcoholic) products out there. It is actually when I look at the professional boards I see a lot of those questions come up. As brewers we are used to having this 1,000 of years in brewing that makes it this stable product that is not overly unhealthy. When you go to a low or nonalcoholic product the things you would normally have like pH and alcohol that keep microbes from growing that are inherently in the process are no longer there. They are finding that they don't have that so they end up having to buy pasteurizers and then they have to figure out which pasteurizer should you be buying depending on what product you are making. I see that all over those boards right now. Hop water has been pretty popular as well.

Pasteurization
One way that a brewer can extend the shelf-life of their beer, including nonalcohol varieties, is to pasteurize it. Advantages include killing spoilage microbes, preventing yeast growth, reducing the need to refrigerate and expanded market distances. Disadvantages are primarily associated with taste and early staling. Batch and tunnel pasteurization are common methods used in beer production. The temperature used to pasteurize beer will depend on the method being used, with the end goal being to raise the beer temperature to approximately 140 °F.
https://prowm.com/pasteurization/methods-beer-beverages-tunnel-batch/
08/10/23

20.5 Professional affiliations

20.5.1 What current professional organizations are you affiliated with? Why did you join them?

I am a member of the MBAA and currently the scholarship chair for them. I am also a member of ASBC and a member of the Pink Boots Society. I knew about Pink Boots before I even got a job in the industry. I knew about Teri Fahrendorf and what she was doing was very awesome and so that was one of the first things I did when I got into the industry. After that being in QA/QC it wasn't hard to convince my brewery to pay for ASBC because of the value that their Methods of Analysis offer members. Then I went to the Craft Brewers Conference, which is more of a trade show. I went to my first MBAA meeting and was like this is where I really wanna be. It was all technical and science based. I've been to the Brewing Summit and Brewing Conference several times. It is a perfect place to live for the size of my brewery. I think once you get larger you get more benefit out of the ASBC.

20.5.2 Could you fill everyone in on Pink Boots Society and what they are about?

It was initially started by Teri Fahrendorf. She had left her brewing position and I believe her and her husband got into an RV and started traveling the country. She stopped at breweries where women were brewing and she realized that she was not on an island and the people she visited also realized that they were not on an island. So they started this network that eventually started Pink Boots. The main mission of the society is to empower women in the brewing industry. Specifically by providing opportunities primarily by education to help them get into the industry. I believe there are four scholarships a month now that women can help with their continuing education to help them move up in their positions.

Pink Boots Society
According to their website, Pink Boots "aims to assist, inspire and encourage women and non-binary individuals in the fermented/alcoholic beverage industry to advance their careers through education." Resources available to members include scholarships, forums and volunteer opportunities. Chapters are located across the United States as well as several countries, including Australia, France and the Netherlands.
https://www.pinkbootssociety.org/
09/01/23

20.5.3 What do you enjoy most about going to their events?

Pink Boots is primarily chapter based, unlike MBAA, that is one big conference. What I like about MBAA and ASBC is the scientific and technical aspects of it that really appeal to me. You see the trends that everyone is talking about or see the research that everyone is doing.

20.5.4 How important is it to be associated with other professional organizations?

I find that not only is the networking aspect pretty awesome but when you have a problem you have people that you can go to. I've made connections with people who are smaller and bigger than my breweries and sometimes make the same product. So for example, how do you coffee that porter, what trials and tribulations have you gone through? We are about to start and if you could help that would be great.

20.6 Other/miscellaneous

20.6.1 Outside of brewing what other hobbies or interests do you have?

I dance with a group of friends. We have quite a bit of dance styles. I am a huge cinema-phile so I watch a lot of film. I have three dogs that I hang out with all the time – pocket pit, lab mastiff mix and a Shepard-Rottweiler mix. They are all pretty hardy dogs and they need a lot of attention. A big dog lover.

20.6.2 Is there anything else that you'd like to share about your personal or professional life that you think people might be interested in knowing?

Not really. It seems like a weird transition to go from anthropology to beer but if you hang out with enough anthropologists you realize that it is not a strange transition. Everything that I learned from my STEM career has really helped. Brewery owners and managers like to hire people who have confidence and know how to do that trouble shooting and how to use the scientific process and keep an eye on everything.

21 Drew Russey

Texas A&M, Zoology BS (2008)

University of Houston, PhD Biology (2014)

St. Arnold's Brewery Houston, TX, QA/QC Director

Texas A&M University https://www.tamu.edu/ 08/21/23	University of Houston https://uh.edu/ 08/21/23	St Arnold's Brewery https://www.saintarnold.com/ 08/21/23

21.1 College/degree pursuit

21.1.1 What inspired you to pursue a STEM degree?

I originally attended Texas A&M because I wanted to be a veterinarian. They have a very good vet school and after a couple of years I decided that vet school wasn't for me. That was about the same time that I started doing undergraduate research, and I enjoyed working in a research lab.

21.1.2 How long did it take you to graduate? Were there any stumbling blocks you'd like to share?

It took me 4 years to graduate. I did not have any real stumbling blocks except when I went to a career fair, and nobody knew what a zoology major was or what they actually did. I went up to the tents that were looking to hire biology majors and I had to explain that it was the same degree except for two different courses.

21.1.3 Did you do any research as an undergraduate? If so, what was the project?

I was tasked with developing a teaching lab for an Animal Physiology lab course. I was looking at how epinephrine, adrenaline and other hormones affected coloration in red drum fish. Basically, I would dose fish scales with hormones, and then count

https://doi.org/10.1515/9783110798777-021

and stage melanophore cells on fish scales. Basically, when they are stressed the cells aggregate and become tight little dots and when the fish are relaxed the cell expands. This is interesting on a cellular level, but what was interesting for me was the fish on a macro level would pale or darken in color in these areas.

Melanophores
"Melanophores regulate melanosome trafficking on cytoskeletal filaments to generate a range of striking chromatic patterns. The mechanism of physiological color change by these melanophores encompasses both physical and biochemical aspects of melanosome dynamics."
https://www.intechopen.com/chapters/37877
08/21/23

21.1.4 You mentioned doing a PhD. Did you do that because you wanted to get into teaching? What was that research work in?

Yeah, developing the teaching lab was my first experience teaching and I really liked it. I had a really good relationship with a couple of my professors and thought I could see myself in a similar role. I pursued a PhD originally planning to be a professor. For my research, I studied phenotypic plasticity and morphological evolution using fruit flies as a model. The size of fruit flies, and many ectotherms, is highly influenced by the temperature of development. Basically, I was looking at how temperature affected fruit fly morphology and body plans, and then how that affected their flight performance and ultimately their survivorship in presence of predators.

21.1.5 Do you have any words of wisdom for students wishing to pursue a STEM degree?

I had an opportunity to give a keynote at a UH graduate student symposium in the Spring. I told them a science degree is great because you can find a job in essentially any field, because science is everywhere. With a STEM degree, you can do quality control, R&D, or something along those lines, because you use many of the same principles. Just look at where your hobbies are and figure out how science can be applied toward them whether that is analytics, food and beverage, or any other area.

21.1.6 How often do you use what you learned as an undergraduate in your current job?

Honestly, it's hard to point to specific course content. What I use the most is the scientific method and experimental design as my approach to answering research questions. It is more about the process as opposed to any specific course knowledge gained. However, maybe I am discounting background knowledge of basic botany and microbiology gained through labs and coursework. There is that basic understanding of plants, fungi and bacteria. I could not point to a particular class though. I never took a brewing course as an undergrad.

21.2 Movement to brewing field

21.2.1 What was your "entry beer" into enjoying craft beer? Why did you like it?

I enjoyed the darker beers such as Shiner Bock and Guinness. Those led me to Samuel Smith's Nut Brown ale. I really like those darker British styles. My first craft beers I got into were Left Hand Milk Stout and Saint Arnold Amber Ale. I began homebrewing around the same time I was getting into craft beer. I started off brewing clones, but I eventually found that if I was spending a Saturday making beer I wanted to make something that I could not just buy at the store. We don't have as many dark beer options in Houston, so I homebrewed typically darker styles. I like the malty, malt forward bodies that are a little sweeter and less bitter. I do prefer the balance of those.

St. Arnold Amber Ale
This 6.2% ABV beer is described as a "well balanced, full flavored, amber ale. It has a rich, malty body with a pleasant caramel character derived from a specialty Caravienne malt. A complex hop aroma, with a hint of floral and citrus comes from a combination of Cascades and Liberty hops."
https://www.saintarnold.com/year-round-beers/
08/21/23

21.2.2 Did you get started with homebrewing? What was your first recipe?

My wife got me started with a basic homebrew kit for Christmas 1 year that included a Samuel Smith's Nut Brown clone.

Samuel Smith Nut Brown
Glad we could include this one in the book. The unique aspect of this beer is it is fermented in Yorkshire Squares. This 5.0% ABV beer is described as being "a relatively dry ale with rich nutty colour and palate of beech nuts, almonds and walnuts." The second link has a great picture of the Yorkshire squares and describes the fermentation process using these.
https://samuelsmithsbrewery.co.uk/shop/bottles/ale/nut-brown-ale/
https://samuelsmithsbrewery.co.uk/the-brewery/
08/21/23

Nut Brown   Yorkshire
             Squares

21.2.3 Did you ever enter a recipe into a brewing contest?

No, I never really did. I brewed for fun, it was never a competition for me.

21.2.4 What was your biggest disaster?

Bottle bombs. Bottles popping off in a closet were my biggest disaster. I kegged my first batch soon after.

21.2.5 Were you part of a homebrew club? Which one? How long a member?

I had friends that I would brew with, but not a formal club.

21.2.6 What was your biggest batch as a homebrewer?

I only have a 5-gallon setup, but I have done 10 gallons by getting two batches done in one weekend.

21.2.7 Did you ever brew a beer for an event (wedding)? How did it go?

I had a friend that would do an Oktoberfest party every year. I was in charge of brewing the non-Oktoberfest beer for attendees who didn't want a "dark" beer. So, I made a lime lager for it.

21.2.8 Prior to getting into brewing what else did you do after graduating?

I actually went straight into the industry. I had one weekend between when I defended my PhD dissertation and when I started at St. Arnolds. I was working on my dissertation edits the first couple of weeks there so it was a hectic two weeks.

21.2.9 Would you recommend brewing to other STEM majors out there and why?

Yeah. There are many different ways to apply a STEM background in the industry. In smaller breweries or nanobreweries, it's going to be similar to opening a restaurant where you are serving the local community. You will wear a lot of hats and need to be flexible day to day and thinking critically on the fly. Then you can go all the way up to like Sierra Nevada or even larger corporations like InBev where you will have a more narrowly focused laboratory job. There are opportunities to do full on research to actual degree programs like at Oregon State. All our malt and hop suppliers have quality control labs as well. Improving and maintaining quality always needs to be done so breweries are always looking for people to do that. We need people to look at why beers are turning out a particular way. It is a fun, friendly industry and a great chance to put your problem-solving skills to work. The industry is willing to help each other solve problems.

21.3 Current position

21.3.1 What are your current job responsibilities and tasks?

I am currently the laboratory manager at Saint Arnold Brewing Company. I manage our production data, track fermentation trends, and analyze beer trials, but I also handle day to day activities like laboratory inventory and scheduling. We track data on everything we brew, everything that is fermenting, and everything we filter. The metrics are endless. I am in charge of integrating our data from paper forms to excel and then to more formalized QA/QC software. The lab essentially makes sure that fermentations are on track and avoid surprises, because surprises are typically bad. This includes trends in our yeast viability, or changes to IBUs when we change our hop lots.

21.3.2 What classes that you took undergraduate or graduate-wise apply to some of the things that you have done in the brewing industry? Why/how?

I would say the most important ones were statistics and anything that had an emphasis on coding. I do a lot of coding and statistics. I did not appreciate the utility of these courses when I was going to graduate school, and I wish I had more formal training in these areas. My advisor would not splurge on a JMP account so I had to do everything in R, which now I really appreciate. I would say that outside of that it was more about the research experience than any particular course. I did not take a full-on botany course, which in hindsight may have been useful jumpstarting my understanding of hops and the grain itself. Much of my coursework was more animal focused than yeast or the like. From that standpoint, some of the evolution course work has been helpful for when it comes to understanding yeast health and yeast management. When tracking yeast health, you are basically monitoring selection on a large population of cells. Yeast is sometimes treated as just another ingredient, but that is far from the truth. They are just trying to survive and end up in a sealed vessel of their own waste. Yeast are probably one of the more frustrating parts of beer, because yeast could be doing fine and then all of a sudden they quit on you and you have to figure out why they did.

21.3.3 Want to revisit the coding work. What would you say is the most common coding language in the brewing industry?

I would say that R is, at least in my circle. Partly, because it is open source and craft breweries often just don't have the funds to buy an expensive software. Anything that is shareable is very helpful. There are all sorts of packages and tools to help you. I do know some that use python and if you are involved in marketing, you may use html or JavaScript for building a website. I use R every day for statistics.

R coding language
R is described as "a language and environment for statistical computing and graphics."
A good thing about R is that it is free and that it can run on many different platforms.
https://www.r-project.org/about.html
08/21/23

21.3.4 What adjustments did you need to make from working in a lab setting to a brewery setting?

The biggest change is balancing production with experimental design. You can't always do the optimal experimental design and instead need to fit your trials to the production schedule. There is the opportunity for benchtop trials of course, and often that's necessary before scaling up. A long time ago we had a winter stout and I had some pH adjustments that I wanted to make, but we stopped brewing that one and the trials were then unnecessary. So, you must keep an eye on the production schedule and design from there.

21.3.5 What do you enjoy most about your current position?

I like the collaborative environment and it's been my experience that it is a friendly company within a friendly industry. It is just a fun environment to work in.

21.3.6 Do you have any words of wisdom for students wishing to apply their STEM degree toward getting into the brewing industry?

You need to be persistent. There are only so many lab specific jobs if that is what you are seeking. If you are applying to a smaller brewery, you may end up wearing a lot of hats before you can focus on the lab work. You need to apply quite a bit and be flexible.

21.4 Experiences in the brewing industry

21.4.1 How would you describe your leadership style?

My philosophy is if you hire the right people then you should not have to micromanage. That is not to say I don't keep up with what people are doing, but you shouldn't need to set an employee's agenda every day. I want our lab techs to have a routine and when we need to do something outside of the routine then we talk and figure out what needs to be done. I also try to hire complementary skill sets. For example, chemistry is not my strong suit, so we hired chemists to help compliment that. We did not have anyone with that background on our team before that and I have learned a lot from them.

21.4.2 What are the various ways that you keep yourself grounded and you use to take care of yourself?

That is not usually a problem for me. I feel like it's something we joke about here. With the brewers in the lab, we come from many different backgrounds, and we are all just basically working at the same place on the same plane. We have people in production from the music industry, people with advanced degrees, people from the armed forces, but very few that went to school specifically to brew. We all have the same goal to make great beer. As far as taking care of myself, I think enjoying your day to day goes a long way.

21.4.3 Do you have any advice that you would give to your younger self before getting into the brewing industry?

I probably would have applied sooner. I did not need a PhD to do most of the stuff that I'm doing. I was probably over-qualified when applying in that regard. I finally applied when I was lining up a post-doc with a wine lab that worked with yeast and thought "I'm only doing this so that I can get into the industry. Why don't I just apply now?"

21.4.4 Do you have any tips or tricks for maintaining a positive work-life balance?

If you enjoy your job that helps quite a bit. Every job has bad days, but I still look forward to what I am doing. I think when life feels unbalanced, it's usually because you do not like what you are doing either personally or professionally. For me personally, the 24 h production schedule allows me pretty good flexibility in work hours since there is always something going on. This has allowed me to come in early, so I can leave early to pick kids up from school and attend after school activities. It saved us a lot of money when the kids were in day care too.

21.4.5 Describe your toughest day in the brewing industry so far

It's hard to pinpoint a day, but it is tough when we have a quality issue come up severe enough for a dumped batch. I am often the bearer of bad news, and it sometimes feels like my fault even if I just happened to be the person who found it. Anytime we have product loss, nobody is very happy.

21.4.6 St. Arnolds has grown quite a bit in the past few years, what challenges are associated with that?

We were already in this building when I started, so I was not here for the transition from the old to the new location. We've seen the barrelage increase, but I'm probably not the best to answer that particular question. We have expanded into more products though. We have been brewing cider for a few years and had a brief entry into seltzers. Anytime we branch out into a new product there is a ton of research into what we need.

Barrels

A barrel, of course, is 31 US gallons. In 2022, close to 25 million barrels of beer were produced which is impressive, but does mark a ~3% drop in total beer production compared to 2021.

https://www.brewersassociation.org/statistics-and-data/national-beer-stats/

08/21/22

21.4.7 Do you want to talk about some of the MBAA workshops and your 2018 MBAA article?

Yeah, it's been fascinating. I really didn't anticipate the response we got and the reception to the paper definitely gave me the confidence to do other presentations like those with MBAA and ASBC. Since then, I've helped with a few workshops focusing on handling brewhouse and cellar data with the aim of helping others analyze similar projects. We are constantly working to improve our beer quality, and if we come across something that is worth presenting or publishing then we will do it. We are a production facility first though, so peer-reviewed articles are not necessarily our priority. However, we have been very supportive of presentations, especially in the local community.

21.5 Professional affiliations

21.5.1 What current professional organizations are you affiliated with? Why did you join them?

I am currently a member of MBAA, but I am most active with ASBC. I serve on the Technical Committee as Chair of the Craft Beer Subcommittee. The Technical Committee oversees all the Methods of Analysis which serves breweries of all sizes. Since anyone can be a member, the Craft Beer Subcommittee makes sure that the ASBC is still developing tools and methods that the craft breweries can use, and not just methods involving HPLC and GC-MS/MS – instruments that are too expensive for most craft brewers. Most breweries do not have that type of equipment. For example, the Chairs before me made the fishbone diagrams used in troubleshooting several common issues. Currently, we are working on spreadsheets to count yeast or monitor fermentations. When we are giving tours to breweries we are often asked where we got our brew and cellar sheets, so our goal is to provide a starting template. Everyone will favor tracking one thing over another, and no form will be perfect for everyone, but hopefully it's a start. We are building these templates in Excel to keep them accessible. I'm sure the next Chair will likely take us in a different direction. We are trying to make these spreadsheets as straightforward as possible without macros or logic checks. We want to keep it simple. We plan to tie them together into a simple dashboard.

Fishbone diagrams
The ASBC has generated a variety of fishbone diagrams that address a number of topics (33 so far), including: yeast management, many process controls and beer stability. These valuable resources are only available to ASBC members and apparently is reaching 200 different fishbones.
http://fishbones.asbcnet.org/
08/21/23

21.5.2 What do you enjoy most about going to their events?

I enjoy the ASBC conferences because they feel the most like academic conferences, but with beer. It's interesting to see what other people are doing, and the cool research that is out there. It is great to have something you can take home that is current and addresses what issues are on the horizon and learn new tips and tricks.

21.5.3 How important is it to be associated with other professional organizations?

It is helpful. I'd recommend attending those meetings even before having an affiliation, because you will see people from different areas of industry. If there are specific topics that you are interested in you can have conversations with those people that are doing that work. At least with ASBC there are professors that you can have conversations with and meet some of their graduate students if you are interested in graduate school. You can also talk to people at breweries who make beers you enjoy.

21.6 Other/miscellaneous

21.6.1 Outside of brewing what other hobbies or interests do you have?

Pretty much sports and board games. I still follow college football pretty closely, mostly University of Houston and Texas A&M, but I'll watch games all Saturday when possible. Since I was born in Ohio my brother and I are Browns fans in addition to being Texans fans. One of these days I will cheer for a team that is successful. That said, the Astros run has been fun, especially since our brewery is located so close to the stadium. I really enjoy science themed board games like Photosynthesis and Evolution and they're a fun way to introduce science to my kids.

Evolution
Evolution is board game made by North Star Game Studio. It allows players to generate their own species using 129 different trait cards. The manufacturer boasts that over 12,000 combinations are possible as a result.
https://www.northstargames.com/products/evolution
08/21/23

21.6.2 Is there anything else that you'd like to share about your personal or professional life that you think people might be interested in knowing?

I don't think I have anything profound. Though, I do think people are often surprised how much science there is behind brewing.

22 Riley Wertenberger

BS University of Kansas (2012), Microbiology; Also ran Cross Country and Track

Alma Mader Brewing, Brewer and Leader of Packaging and Quality Assurance

University of Kansas
https://www.ku.edu/
09/01/23

Alma Mader Brewing
https://www.almamader
brewing.com/
09/01/23

22.1 College/degree pursuit

22.1.1 What inspired you to pursue a STEM degree?

I started out in engineering actually. I initially wanted to do chemistry because I enjoyed chemistry in high school, but I thought engineering was more practical so I could make a lot of money coming out of college. Engineering turned out to not be very interesting to me. I was also doing sports. I ended up taking a microbiology class the beginning of my sophomore year and really got into molecular biotechnology and just how proteins interacted. I also really enjoyed organic chemistry. So I decided to switch my major to microbiology. At Kansas University it is all medical. If I knew that if I was going to go into food science I probably would have gone to Kansas State or somewhere else that specialized in that particular field. KU was mostly medical based microbiology and human physiology.

22.1.2 How long did it take you to graduate? Were there any stumbling blocks you'd like to share?

It took me 5 years to graduate. I think my freshman year I took classes all year, was doing athletics and then took courses during the summer. I was extremely burned out and did poorly during my sophomore year. I actually had to change majors so that I could stay eligible for sports. I think it's important for every student to give themselves a breather, maybe take summers off. My brother had a similar experience. Bad grades happen, accept it and keep moving on.

https://doi.org/10.1515/9783110798777-022

22.1.3 Did you do any research as an undergraduate?

I did. The main take away that I got from it was just working in a lab. As an undergraduate you get credit for taking a lab, but it is different with research. You are learning how a lab functions, how to order consumables, how to work different machinery. I learned how to do PCR, how to use a nanodrop and how to use a micropipettor. You know, some of the same things you are learning in class, but it is just a higher up knowledge. I'd really recommend undergraduate research.

Nanodrop

These are really handy when you are working with small volumes. They only need
0.5–2 μL of liquid (model dependent) for analysis and can measure DNA, RNA, protein
just to name a few.

https://www.fishersci.com/shop/products/nanodrop-one-spectrophotometer/13400518
09/01/23

22.1.4 Do you have any words of wisdom for students wishing to pursue a STEM degree?

They are really fun careers. I have never made a lot of money in my particular line of work. I have done undergraduate research, I've been a research assistant, I've done other molecular bioscience work. I really have enjoyed my career in the brewing industry and have found out how scientific and creative it can be. I have truly enjoyed going into work every day. I would say to find that part of science that lights you up and do it.

22.1.5 How often do you use what you learned as an undergraduate in your current job?

I would say that currently in my job I use my math and chemistry skills quite a bit. I used it a lot more at my previous brewing jobs. Right now I am a brewer, but I'm also in charge of QA/QC and I am the lead of packaging. There are only three full time employees. Two of us are in brewing and production and the other one does payroll, HR and everything else. Before being here I was at Boulevard Brewing Company from 2015–2018. I was their microbiologist and put together their PCR program. I also turned the micro department around to make sure we were using our money and time better. I also made sure that we were being more smart about sampling and people knew how to do it right. From 2018 to 2020 I was at a brewery in California as their quality manager. I supervised three individuals – a microbiologist, a chemist and someone who did sensory and packaging things for us. I came up with the whole quality scheme for them. I used chemistry and biology courses in particular. You need to understand yeast and how it works and what it does and does not like. It is really like using what you learn in college as a basis to understand and learn more about the art of brewing.

Boulevard Brewing
The beer I remember from this brewery is the Bourbon Barrel Quad. So many great tasting notes in that one.
https://www.boulevard.com/
09/01/23

22.2 Movement to brewing field

22.2.1 What was your "entry beer" into enjoying craft beer? Why did you like it?

Wilds and sours were popular when I got into La Folie by New Belgium. That blew my mind. I was starting to understand how much yeast can influence beer and its taste and aroma. I did not know much about it in college. I ran and did not drink much in college. I went to some beerfests after graduating and was thinking, this is really cool. I remember having a couple of beers and talking to someone at the booths. I walked into a microbiologist and talked to them for a couple of hours and was intrigued by the esters and the phenols and all the other aromatics that yeast can contribute to beer. Ale versus lager yeast and their attenuation rates were interesting to. Also I like that the wild yeast and lactic acid bacteria can give it that desirable sour quality.

La Folie New Belgium
"Eccentric Madness": This beer is a "flanders-style sour brown" that is aged in huge foeders. La Folie is described as having a "sharp and sour flavor, full of green apple, cherry, and plum-skin notes."
https://www.newbelgium.com/beer/la-folie/
09/01/23

22.2.2 Did you get started with homebrewing?

No. I have never homebrewed. I watched a friend do it and of course it was all about cleaning. I was very interested in the process though and did a lot of research. I read a book about brewing yeast. I read some books about barrel aging. There was a malt, yeast, water hops series and I thought it was really cool. I was doing PCR for a pharmaceutical lab and they were doing a lot of drug testing. So, I was pretty bored with it and at the time did not see a future in it. I bugged Boulevard brewing for about a year to see if I could get an internship. They were actually looking for somebody to start a

PCR lab in their laboratory. I got an internship with them and then about a year after that they were looking for a microbiologist and so I got that job.

22.2.3 Prior to getting into brewing what else did you do after graduating?

After I graduated I worked in Kansas City. I was a research assistant there in a molecular bioscience lab for 2 years. I was working on a project on uterine fibroids. We were raising mice with a genotype. I got to do some cool things like splicing genes and doing Western blots and things like that. It wasn't my most favorite job ever. I did learn a lot of biochemical procedures and things like that. After that I went to a place that did pharmaceutical research and I basically did PCR all day long for them.

Western Blots
These wonderful tools allow you to visualize protein components from a sample. Proteins are separated using gel electrophoresis, then transferred to a membrane and finally the target protein(s) are visualized using tagged antibodies.
https://www.ncbi.nlm.nih.gov/pmc/articles/PMC3456489/
09/01/23

22.2.4 Would you recommend brewing to other science majors out there and why?

Absolutely. I love my hob. Right now I am in a unique position and I started in science. Most brewers out there just get started in brewing and then they learn that there's this real cool science part to it. My background is in quality assurance. I took a two week course from Siebel in Montreal and learned microbiology. It wasn't necessary, but I learned a lot there. I have a unique perspective since I have a science background and I've been taught progressively more and more about production work.

I think if you enjoy getting your hands dirty and working really hard and in a fast-paced environment you would enjoy it. It is very physical and you are like using your mind a lot. In my particular positon you are wearing many hats every single day. It is so satisfying and you can be very rounded out and get familiar with the whole process from raw ingredients to the end product to how the yeast is interacting with the beer and ingredients. There is quality assurance all the way through receiving your raw materials, using them, managing them all the way up to packaging. Even selling the beer, keeping it aseptic and making sure you get the right carbonation and you are trying to prevent oxidation and making sure that everything is up to specs. All of that is definitely science.

Siebel Microbiology course
We've mentioned Siebel in a couple of chapters, but wanted to highlight the brewing microbiology course here. It is a wonderful tool for those who are looking to build a quality program or are lacking the formal training in this area.
https://www.siebelinstitute.com/courses/intermediate-level/brewing-microbiology
09/01/23

22.3 Current position

22.3.1 What are your current job responsibilities and tasks?

My current title is brewer and leader of packaging and quality assurance. Like I said there is one other person in production with me and so we pretty much do everything. My job is I brew the beer and I get to work at 7 AM and brew all day. We make sure we are hitting specs throughout the day and assuring that everything is cleaned up. We like to keep the brewhouse and the cellar very clean and so I get in, sanitize tanks and transfer brew into those. We then clean out spent grain and assure that things are clean. Other days it is cellar work and I'm taking gravities, temperatures and pH every day. While we're fermenting I take samples before we go to crashing. We test for VDKs and diacetyl to make sure that it is completely done before we go to crashing. When the beer needs to be transferred to a bright tank I make sure our tanks are cleaned and purged. We also do 16 ounce cans and 500 mL glass bottles. I will prep the canning and glass lines and make sure that they are clean. We will then package all day and get the kegs ready. We make sure the hot and cold liquor tanks are ready to go. I think the most important thing about a brewery is to keep it clean. It can get dirty very easily and contamination is a huge risk. I did a seminar in DC last year. It was about breweries who keep a wild and regular yeast in the same building and how you keep it separate. We do ciders once a year and so we want to keep everything clean. When we are racking out of barrels I will take samples for micro and do cell counts before pitching. If something is concerning then I'll take a look at it in the lab. We also want to make sure that we don't over or under pitch. Similarly, we do not have cans to work with so testing for contamination is not a huge risk for us compared to breweries that do can.

22.3.2 What is your favorite and least favorite analysis to do and why?

My favorite one is to do VDK analysis. When I was in California I came up with another method to analyze VDKs using a distillation method. ASBC has a method for doing this and I kind of took the method and ran with it a little bit. I think everyone needs to tweak those for their method. I really enjoyed doing method development and testing it out. My least favorite one would probably be pouring plates. I don't like doing that too much.

22.3.3 What classes that you took apply to some of the things that you have done in the brewing industry? Why/how?

Chemistry for sure and learning how to do stoichiometry. I use that almost every week. Microbiology really helps you to learn how proteins function. I would also say immunology because there are a lot of interactions with other materials, but not human. Just understanding how your yeast work and whether they are killer yeast and the like. Also, how they interact with the environment around them. There are so many different aids that people use and if you don't understand how the yeast are using it then it could be that you are solving one problem but maybe it is also giving off another flavor that you do not want. It is important to understand how proteins are interacting with compounds and what the end results are. If you want to open up your own brewery then physics is very important as well.

22.3.4 What adjustments did you need to make from working in a lab setting to a brewery setting?

It was actually easier to do a brewery setting. Academic lab work is very solo and based on self-motivation which is totally fine. In a brewery it is quite a bit of teamwork. There are so many moving parts and it is extremely fast-paced. I think that might be the most difficult thing to get used to. There may be things where the quality manager is saying, "Here's the problem, here is how we might be able to fix it" while the sales team and the brewmaster is like "okay, but we have to sell this beer." It is sometimes a go no-go, but you have to fix the problem so it can be sold. The hardest thing is, I do not want to say getting over the ego, but you still have to get the problem solved in real-time. You can make sure that it does not get repeated in the future. There is a whole lot of problem solving and creating procedures for the future to make sure that everyone's training is where it needs to be. The fast paced is definitely the biggest difference.

22.3.5 What do you enjoy most about your current position?

I like that it is really active. The people are really great to work with. Most of the people I know in brewing are excited to learn. They want to know more about brewing and make really good beer. I think it is really just the community as we. We have a community around us that supports us and we talk about beer and they care about how you are doing. There is also a community of breweries and I guess I'm just talking about Kansas City right now. If something happens or you're short of a bag of grain your neighbor probably has it. We have a few breweries we work with in the area that we bounce ideas off of, or share raw materials needed. We want others to succeed for sure and want to push each other to make really good beer so the community is awesome.

22.4 Experiences in the brewing industry

22.4.1 How would you describe your leadership style?

I definitely lead by example. I am a really hard worker and may not know the answer immediately, but I am willing to do the work to find out. I just want everybody to feel empowered. I like training people so that they have the keys to move forward which is very important.

22.4.2 What are some unique strengths that you think women bring to the brewing industry?

I think we give a little different perspective. There is a different type of communication style that they bring to the industry. I hate to say this, but a while back craft brewing was more of a boys club and it was uncomfortable for women to enter the industry. There are a lot of women who persevered and showed their strengths. There is a different perspective and the understanding that women also like to drink beer and be a part of the community where beer brings people together.

22.4.3 Have you had any negative experiences in the brewing industry that you would like to share?

There has been a huge improvement in the last couple of years. There are experiences regarding harassment and degrading comments toward women. My first couple of years was kind of all over the place in regard to those experiences. I think it is very important for the industry and the service industry to realize that women are a definite part of this now. I do not know another women in the industry that has not at one time experienced this. I remember when someone came into the lab and was talking to my chemist. I was the lab manager and they kept talking to the male chemist. I finally told them that they were selling things to me and not them so they either had to start talking to me or leave. It takes a willingness to stick up for yourself. Be nice, but still stick up for yourself. There are some exceptional women who care about beer and want to make good beer. It is definitely improving which is good and I really like the environment that I am working in now.

22.4.4 Do you have any advice that you would give to your younger self before getting into the brewing industry?

I did not understand the culture when I first got into the brewing industry. I have brothers and uncles who toughened me up. When I got into the industry there was a lot of pushing me to the limit to see how much I could take. I would have told myself to stand my ground and know your values. I would have told myself to get into the science and be kind.

22.4.5 What are the various ways that you keep yourself grounded and you use to take care of yourself?

I make sure that I run almost every day and make running and food prep a priority. It helps to keep myself sane and healthy. If I do not get a run in everything just spirals downward. I am at work 60 h a week and making sure I get a workout in everyday and that I am just actually eating. I make sure that I spend Sunday with family and friends and go to church and be myself.

22.4.6 Describe your toughest day in the brewing industry so far

There's actually a couple. My toughest day that stands out is when I first started at Boulevard. I used to hydrate dry yeast for our program down the road. We would hydrate it in 5 liter carboys. I added the yeast to the water and did not realize that I tightened the lid too much. There was a ton of backpressure on it. I tried to lightly loosen the lid and to relieve the pressure and the lid blew off and yeast was all over the lab-ceiling, microscope all over the place. A coworker and I spent like three hours cleaning everything up. That was a very tough day. There are going to be days like that and at times it will be an overwhelming job, but just keep learning.

Yeast brink
A yeast brink is usually much larger than the 5 L vessel that Riley mentioned. These have inlet and outlet ports, a thermometer and wheels so they can be moved into place. Their purpose, of course, is to allow brewers to harvest and move yeast between fermenter. https://www.craftmasterstainless.com/yeast-brinks
09/01/23

22.5 Professional affiliations

22.5.1 What current professional organizations are you affiliated with? Why did you join them?

I am a member of ASBC since 2016 and there are so many great standard procedures (Methods of Analysis). I was at the New Orleans meeting which was cool. I like access to all the scientists in the field and all of the new procedures coming out. It is great information to have. I'm also associated with MBAA and am trying to become a board member for the MBAA in the Great Plains. It covers Southern & Central Iowa, Kansas, Western Missouri, Nebraska, and Oklahoma.

22.5.2 What do you enjoy most about going to their events?

I like getting to know other people in the industry. There are several people that I met in different labs. I know the guy whose doing Creature Comforts and we met at ASBC. We worked on a genotyping project and they were surveying everyone's Wild Strain house yeast. I think it's neat to meet people at other breweries and learn what they are interested in. If you are ever traveling you can talk to people at their brewery and learn about their operations.

Creature Comforts
Talk about a brewery on a mission, "We exist to foster human connection. Our hope is to be a force of good in the world through development of industry-leading beverages and experiences." They have several impact programs they work on.
https://creaturecomfortsbeer.com/
09/01/23

22.5.3 How important is it to be associated with other professional organizations?

It's very important because in brewing it is who you know. If you want to move your career forward there is a community of people who can help you with the next step. If you're looking to move as well it gives you a great chance to have a long career in science.

22.6 Other/miscellaneous

22.6.1 Outside of brewing what other hobbies or interests do you have?

I run quite a bit and do marathons. I also enjoy cooking and baking. I enjoy gardening and have quite a few plants in my apartment right now. I also enjoy music and movies. I have a Golden Doodle named Penny and we go to the dog parks quite a bit. I also like being outside and being with my friends. I also enjoy traveling. I just did the trail to Machu Picchu last year.

Marathon training
When I told one of my rugby buddies that my brother was running a marathon his response was, "I'm good. I like to get that many miles in over a month instead of *one* morning." There are many training programs out there to help people reach their goal of running a marathon.
https://www.runnersworld.com/uk/training/marathon/a776459/marathon-training-plans/
09/01/23

22.6.2 Is there anything else that you'd like to share about your personal or professional life that you think people might be interested in knowing?

I think brewing can get glamorized. At times it is very cool and people recognize you as a brewer. It is a difficult job but it is very satisfying if you like seeing people enjoy your hard work. It can be very gratifying. I can't say enough good things about the industry right now.

Section IV: **Other disciplines**

23 Scott Brady

BS In Graphic Design, Arizona State University (2011); Member of ASBC

OHSO Brewery and Distillery, Production Manager

Arizona State
https://www.asu.edu/
08/10/23

OHSO Brewery
https://www.ohsobrewery.com/
08/10/23

23.1 College/degree pursuit

23.1.1 What inspired you to pursue a chemistry (STEM) degree?

It was STEAM rather than STEM for me. I took a gap year off, actually a few, between high school and college. I could not decide between the arts or the sciences. I started with a business major focusing on math but I ended up on the arts route. I figured graphic design was a good way for me to use my creative expressions to build a career. Brewing had a significant influence on me though. My hobby in homebrewing combined the arts and sciences together in a way that I didn't have to choose between the two.

23.1.2 How long did it take you to graduate? Were there any stumbling blocks you'd like to share?

About 5–6 years in total. I went to a community college first and then took a semester off as I started off in the professional world. I had a very casual approach to school. While I loved school it was hard for me to commit to any one thing. I took time to ruminate on how I was to continue. The biggest obstacle was that I was working full-time throughout college while also taking a fulltime credit load. I had to pay for my own living. The only assistance I had was grants and loans. I never wanted to stop going to school but unfortunately, I had to consider the debt that started to escalate quickly once I transitioned to ASU. If I could perpetually go to school without worrying about finances I would do that.

https://doi.org/10.1515/9783110798777-023

23.1.3 Did you do any research as an undergraduate? If so, what was the project?

Not in the scientific area. I did write a few papers concerning the arts.

23.1.4 How often do you use what you learned as an undergraduate in your current job?

In my current position at least weekly if not sometimes daily.

23.2 Movement to brewing field

23.2.1 What was your "entry beer" into enjoying craft beer? Why did you like it?

Sierra Nevada Pale Ale which is a classic. It was one of the few beers that you could get consistently anywhere you went. It was light, hoppy and had that full-bodied flavor to it. Like I said, you could get it anywhere. For so many it was the beer that roped us into this world.

Sierra Nevada Pale Ale
This was a common response among participants. Other aspects of the beer include that it is made with caramelized malts and two-row pale ale. The IBU is 38 and a 12 oz can has 175 calories with 14.3 g of carbohydrates and 1.9 g of protein. "Some things just grip your senses and never let go. They're so intriguing, funny, or delicious that they affix themselves to your very soul. No matter the tides of time, they are Still the One. Like Pale Ale, hopped to perfection and unshakable."
https://sierranevada.com/brews/pale-ale/
08/10/23

23.2.2 Did you get started with homebrewing? What was your first recipe?

I did. The very first beer I made was a Pale Ale because it came with my brewing kit.

23.2.3 Did you ever enter a recipe into a brewing contest?

I entered many competitions.

Beer Judge Certification Program
The Beer Judge Certification Program website is where most brewing competitions get their guidelines for judging beers. It is an excellent resource for individuals wishing to become a certified beer judge and includes information on style guidelines, education and how to set up competitions. As of August 2023, there are over 8,000 active judges with over 12,000 sanctioned competitions and over 2 million beers judged!
https://www.bjcp.org/
08/10/23

23.2.4 What was your biggest disaster?

As a homebrewer the biggest one was probably around that fourth or fifth batch I ever did. It was a sanitation problem. We were making a brown ale and we ended up making malt vinegar. It was very disappointing but lessons were definitely learned.

23.2.5 Were you part of a homebrew club? Which one? How long a member?

Yes. I was a member of the Arizona Society of Homebrewers.

Arizona Society of Homebrewers
The mission of the ASH is "To preserve and promote the time-honored tradition of homebrewing, and to recognize it as a true art form through information, education, and dedicated practice." They host three annual festivals: Springfest, Oktoberfest and a Holiday Party and meet on the third Tuesday of the month.
https://azhomebrewers.org/,
08/10/23

23.2.6 What was your biggest batch as a homebrewer?

Ten gallons was my biggest batch.

23.2.7 Did you ever brew a beer for an event (wedding, etc.)? How did it go?

No. I did do brewing demonstrations but never made a batch for a specific event.

23.2.8 Prior to getting into brewing what else did you do after graduating?

I was mostly in the publication sector. I worked for the *New Times* in their art department as I was pursuing my associate's degree. I also did freelancing for a while until I ended up getting a job at a homebrew store. I worked there for about a year until I finally entered into brewing professionally.

23.3 Current position

23.3.1 What are your current job responsibilities and tasks?

Currently, I am the production manager at our main production facility. I oversee all of the operations of the brewery. I have my hands in pretty much everything at this point. I don't brew much anymore but I help others learn and maintain the overall flow.

23.3.2 What is your favorite and least favorite part of the job and why?

My favorite part of the job is the creativity and problem-solving aspect of it. Brewing can get a bit monotonous when you're pumping out the demand for core brands. In brewing, though, there is always something new to try or experiment with. Everything has the potential to improve. One of the reasons that I got into it in the first place is because there is always something new for you to learn and apply. The least favorite part of my job is adjacent to that. When something goes wrong and you're stuck having to fix it on the spot before you can get going with your day. In the end though it never gets boring.

23.3.3 What classes that you took apply to some of the things that you have done in the brewing industry? Why/how?

I used to think I wanted to start a brewery so I started out majoring in business. I'm a huge math nerd and those classes have been helpful. Like algebra is always useful and I actually even end up pulling out calculus sometimes. One of the reasons I enjoy math is once you understand the concepts you can apply them in all sorts of creative ways. Since I ended up switching to the arts, the classes aren't necessarily applied in a traditional sense. Learning the creative arts teaches you to think in creative ways which I think has been the most useful. A lesson I learned once that's always stuck out is the concept of every stroke you make needs a purpose. If you can't justify doing something it likely detracts from the overall piece rather than adding to it. Sometimes something sounds a lot better in concept than in practice, and it's important to remember the big picture without getting attached to trivial details.

23.3.4 What adjustments did you need to make from homebrewing to working in a brewery setting?

When I started working at a brewery it was a very small system. It was pretty much a large homebrewing setup. The initial adjustment was pretty smooth. I would say that overall, the most significant adjustment was putting aside some of my assumptions about beer and keeping my mind open to learning new approaches and techniques. While homebrewing I rarely repeated a batch but in professional brewing consistency is king. One of the biggest challenges in brewing is hitting specs over and over again.

Beer Quality Challenge
Speaking of Specs, the ASBC hosted a 2023 Beer Quality Challenge where breweries are invited to produce a beer, in this case an American Pale Ale, that meets ABV, IBU, SRM and pH specifications. Breweries that submit beers with the designated specifications are provided an ASBC sticker and recognition.
https://www.asbcnet.org/events/Pages/BeerQualityChallenge.aspx
08/10/23

23.3.5 What do you enjoy most about your current position?

There is always something new. I have been to many classes now that I am a professional brewer. I like the creativity aspect of it. With brewing it touches on both the science side of me as well as the creative side. The combination of those two sides has been my favorite part.

23.3.6 Do you have any words of wisdom for students wishing to apply their chemistry (STEM) degree toward getting into the brewing industry?

From a chemistry background you have a lot of opportunity at the larger breweries. Positions for that are more common there than at the smaller breweries. Also, if you have one of those degrees you will likely be taking a pay cut at the smaller breweries. So make sure you are passionate about it. It's different if you're getting into a macrobrewery or a regional one.

23.4 Experiences in the brewing industry

23.4.1 How would you describe your leadership style?

I would say that I am pretty open. I like to allow people to try to problem solve on their own. When you're facing a novel situation it is easy to get flustered and shut down. I like to train people so that they understand why we are doing something as opposed to doing it because it's the way it has always been done. Of course there are always specs and goals you should be aiming for, but everyone has a creative side and letting that come through can often bring to light a better way of doing something.

23.4.2 What are some unique strengths that you think a graphic design background might bring to the brewing industry?

The creative aspect. When you work in creativity you are always working for a client. So you have this box that you have to work in. I like to say that you can't work outside the box until you have a box in the first place. So, it goes back to the creativity and using it to problem solve. That is the biggest thing. If you're designing a recipe you are trying to hit a certain part of the market, how can we make it work for us? If you're designing a brewery you have to figure out how to get all this stuff to fit and make it functional and have a smooth flow to the operation. One of the biggest things in creative endeavors is that you never stop with your first idea–no matter how much you love it. You try to think of as many ideas as you can because you must get the bad ideas out before the good ones show up. Sometimes your first idea ends up being the best but that is the exception.

23.4.3 What are the various ways that you keep yourself grounded and you use to take care of yourself?

I practice mindfulness meditation. Mindfulness has become trendy lately to a point that it just sounds cliché. However, this has been one of the more profound ways I've found to keep myself grounded. It didn't happen overnight but with daily practice it can go a long way. It allows you to see your emotional state in a clearer lens. If something goes wrong and you get upset or angry it's far easier to acknowledge it. Once you acknowledge what's happening inside you, it's easier to move towards a solution. It also does wonders for setting aside your ego to work through interpersonal issues.

23.4.4 Have you had any negative experiences in the brewing industry that you would like to share?

I'd say that the negative ones come from miscommunication. There's lots of moving parts and one getting derailed can send waves of problems. It takes a certain personality type to work in a brewery. We joke that you need a certain level of masochism to work in a brewery. There are also hard days where it is very physical and demanding. Some people just aren't made for it and that's okay. Of course, this is true with many pursuits in life. Often you don't know until you try. Being a male dominated industry there's also sexism. This comes from all sides from customers all the way up to ownership. Luckily there are determined people fighting to change this. The company I work for actively fights this as well which is important to me.

23.4.5 Do you have any advice that you would give to your younger self before getting into the brewing industry?

It's a bit of a hard question to answer as it's been a wonderful experience thus far. I feel like the path I went down had ups and downs but in the end, I learned and grew. In that sense I would simply say trust your instincts and keep your mind open to opportunities.

23.4.6 Do you have any tips or tricks for maintaining a positive work-life balance?

At the end of the day it really is just beer. You can always try again. If something goes wrong you just have to start over sometimes. I try to leave it at work too. Boundaries are important in this sense. For example, with some exceptions, I don't check Slack or Emails when I'm not working. The world doesn't revolve around you. If you're working with a good team and you take care of your end of the bargain, you shouldn't be needed outside of your shifts. Do not work yourself into the ground and make sure you give yourself time to relax. I am a big believer that your productivity and effectiveness take a hit if you're grinding all the time. Taking time off is going to make everyone happier. The beer and the brewery will still be there tomorrow. Lastly, one of the most important aspects is the company's culture. Culture is one of the hardest things to change so if you're not happy its often better to move on.

23.4.7 Describe your toughest day in the brewing industry so far

I'll share one that recently happened. It was pretty tough. We have a distillery that doesn't have much along the lines of fermentation ability. We have been playing with doing fermentations at the brewery and then ship over to the distillery. They wanted to do a rye whiskey. I should have listened to my gut on this because the grist was 50% rye. Rye is very sticky and notorious for stuck mashes. We threw everything at it preemptively to try to counter the rye effects. Inevitably the mash stuck in the worst way. It was so bad it ended up bending the false bottom of our lauter tun. The consistency of that wort was like snot which made for a very messy day. It was tough. We had to go back to the drawing board. We need either new equipment or a new approach if we are going to try this particular recipe again.

23.4.8 Anything else you'd like to share about the particulars of brewing in Arizona or your area?

The biggest thing here is that our tap water is bad for brewing, and so we have a reverse osmosis system that we use to make all of our water and then add back minerals. It gives us a nice level of control. I know that in other areas like Colorado and Asheville their water is great and that's actually part of the reason why some big breweries are there. Related to that, we try to be efficient with our water. With our scale it is hard to do too much. When we built our original facility we did do some things to help with that. A lot of the water is in the cleaning aspect of it so we have a cleaning trolley with two vats. It allows us to reuse the water and heat it back up. I know some breweries do not have that option. We do burst rinsing after we clean tanks. A neat thing we participate in now is Scottsdale's One Water Brewing Showcase. Scottsdale Water has the only system in Arizona that's rated to basically purify waste water to potable. They ship this water to participating breweries to make a beer for their Canal Convergence event. Since wastewater is where most of the water use in a brewery ends up its important that people are making solutions to be able to reuse it instead of sending it down the drain.

Water and wastewater: treatment/reduction manual
This book, produced by the Brewer's Association, is a must-read for individuals who are working at a brewery or thinking about opening one. The manual offers sections on Data Management, Best Practices, On-Site Wastewater Treatment as well as Case Studies on water reduction and treatment.
https://www.brewersassociation.org/attachments/0001/1517/Sustainability_-_Water_Wastewater.pdf
08/10/23

23.5 Professional affiliations

23.5.1 What current professional organizations are you affiliated with? Why did you join them?

ASBC of course. They are great because of their technical articles and procedures. For us it is really easy when we are writing an SOP because ASBC will have a method that is already published and tested. That is a great source of information. They also have their forum boards where we can pick the brains of biologists and chemists. Then they have these events and educational activities like webinars and seminars. There is also MBAA and I've attended a class from them before and they also have many resources like the ASBC. There is also the Brewers Association and their forums. They are great from a safety aspect in that they offer a free courses. There is also our local AZ Brewer's Guild of which I am on the technical committee.

Arizona Craft Brewers Guild
The Arizona Guild recently celebrated their 25th anniversary and serves over 100 breweries in Arizona. Annual events include the Arizona Strong Beer Festival, Brewer's Conference (Brewcon), Baja Beer Festival and an Oktoberfest. Their mission is "To advance the understanding and appreciation of the Arizona craft brewing industry, and maintain the quality and image of beer produced through the education of consumers, retailers, distributors and brewers."
https://chooseazbrews.com/
08/10/23

23.5.2 What do you enjoy most about going to their events?

I've been to a lot of Guild meetings. I have not been to ASBC or MBAA meetings, but I have attended webinars and classes. What I like about the Guild ones is that we go over something good or bad that happened or what you could do better or do their way. They may have a piece of equipment that you can look into how it operates by going there.

23.5.3 How important is it to be associated with other professional organizations?

At least join one of them since they do cost to join. Often times the brewery will have one or they may even offer to pay for one. The information you get from them is invaluable especially when you are making the jump from homebrewing to professional. There are some subtleties that you may not realize.

23.6 Other/miscellaneous

23.6.1 Outside of brewing what other hobbies or interests do you have?

Anything creative. I'm trying to write more currently and pick up some old instruments. I also do stop motion music videos with my wife and a friend. Weather has been a big hobby of mine over the years. I have a cloud themed Instagram and go storm chasing during the Southwest monsoon season.

23.6.2 Is there anything else that you'd like to share about your personal or professional life that you think people might be interested in knowing?

Not really.

24 Will Harrer

B.S. St. John's College, Santa Fe, NM Liberal Arts, 2017

Certificate from Siebel Institute, Intermediate, 2017

Second Street Brewery, Santa Fe, NM, Brewer

St. John's College
https://www.sjc.
edu/santa-fe
08/09/23

Siebel Institute
https://www.siebelinstitute.com/courses/intermediate-level/
wba-concise-course-in-brewing-technology
08/09/23

Second Street Brewery
https://www.second
streetbrewery.com/
08/09/23

24.1 College/degree pursuit

24.1.1 What inspired you to pursue your degree?

The St. John's program is a Great Books program. It is a kind of old school traditional liberal arts program in which there is really only one degree that everybody gets. It has a very small student body of about 300 people and my graduating class was 60 people. It was very philosophy oriented. Very well-rounded, traditional. It also had a very solid backbone in math and science.

St John's Santa Fe
St. John's Santa Fe is a campus of. St. John's College which opened in 1964. The College was founded in 1696 as "King William's School," making it the third oldest college in the United States. The mission statement of the Santa Fe campus states that the College "fosters searching consideration of the most important human questions through reading and discussion of great books." Only one degree (Liberal Arts) is offered.
https://www.sjc.edu/santa-fe
08/09/23

https://doi.org/10.1515/9783110798777-024

24.1.2 How long did it take you to graduate? Were there any stumbling blocks you'd like to share?

It took me 4 years. For me, that is kind of tricky since it is such a philosophically oriented program. In regard to St. Johns, the program is so abstract in its approach to science. A big issue for me was getting lost in the data and losing sight of the bigger picture of why we are asking the questions that we are.

24.1.3 Did you do any research as an undergraduate? If so, what was the project?

I did not.

24.1.4 Do you have any words of wisdom for students wishing to pursue a college degree?

Never stop asking questions and always try to keep in mind why it is that you are there.

24.1.5 How often do you use what you learned as an undergraduate in your current job?

I would say that at a certain level not frequently. On a philosophical level it would be every minute of every day as far as critical thinking, problem solving, learning how to ask the right questions and approach every problem that comes up during the day.

24.2 Movement to brewing field

24.2.1 What was your "entry beer" into enjoying craft beer? Why did you like it?

You'll probably hear this a lot. Sierra Nevada Pale Ale. I probably tried it when I was 16 years old. Before that I had only had your commercial beers that were available. It tasted like trees, it was maltier and probably tasted like something you should enjoy drinking rather than playing beer pong with.

Sierra Nevada Pale Ale
This indeed was a common response when interviewees were asked about their "entry beer." The 5.6% ABV beer was introduced to the market in 1980. It is described as having a "bold hoppiness and smooth malt flavor." Sierra Nevada credits the beer with popularizing cascade hops which are noted for their citrus aroma and pine notes.
https://sierranevada.com/
08/09/23

24.2.2 Did you get started with homebrewing?

I actually started homebrewing when I was a teenager. I brewed my first batch when I was 17. Had a deal with my mom that she would not buy me beer but if I wanted to go ahead and make it then that was totally fine with her.

24.2.3 What was your first recipe?

If you are familiar with Bass, the British Pale Ale. It was a Bass clone from the local brewshop back in California.

24.2.4 Did you ever enter a recipe into a brewing contest?

No, because I would have had to be 21 years old.

24.2.5 What was your biggest disaster?

I tried to make a Rye IPA once. I came up with it early in the morning and I don't know if it was how I designed the recipe but it kind of had the same culture growth that you would find at the bottom of a kombucha.

Rye IPA
According to the Beer Judge Certification Program, the Rye IPA is hop-forward, has a dry finish and clean malt flavor. The ABV range for this style of beer is 5.5–8.0%. A Rye IPA, as compared to an American IPA, will be spicier and drier, but is noted to have a less intense rye malt character than the Roggenbier.
https://www.bjcp.org/style/2015/21/21B/specialty-ipa-rye-ipa/
08/09/23

24.2.6 Were you part of a homebrew club? Which one? How long a member?

No, again because I wasn't 21 yet.

24.2.7 What was your biggest batch as a homebrewer?

Never more than one 5 gallon batch at a time. Only really had one going at a time.

24.2.8 Did you ever brew a beer for an event (wedding)? How did it go?

Nope. Not as a homebrewer.

24.2.9 Prior to getting into brewing what else did you do after graduating?

After I graduated I went back home California to try to figure out what I wanted to do. I worked at my old coffee shop job. I was also a real estate assistant. I figured that I did not like the traffic there anymore. So I moved back to Santa Fe. While I was trying to get a job in the industry I worked as a pedicab driver.

24.2.10 Would you recommend brewing to other science majors out there and why?

I think I would because getting back to the bigger picture. If you really love beer and really love science the brewing industry is a great welcoming environment. You should really get involved in that community and be around that mindset.

24.3 Current position

24.3.1 What are your current job responsibilities and tasks?

Currently I am mostly a brewer. We are small enough, we are growing fairly rapidly though, that most people know how to do pretty much everything in the brewery. I am a brewer I also know how to do cellar work, I know how to operate the chiller. I started off as a delivery driver and just moving kegs around. Eventually, the owner was interested in opening up a QC/QA program. I had quite a bit of interest in that. So, in addition to brewing I am also in charge of their QC/QA program. As the brewer I'm responsible for maintaining all our equipment and continuing plant sanitation. I also handle the actual production of beer from the milling of grains to the actual brewing process and then the cellaring. Then transferring to bright tanks and making sure that it gets into the kegs and cans and cleanly as it should be.

24.3.2 What classes that you took apply to some of the things that you have done in the brewing industry? Why/how?

I would say because the science and math are all kind of grouped together. A lot of the eighteenth- and nineteenth-century physics as well as the microbiology. One that comes to mind is carbonating beer. There is a lot of physics that goes into that. Knowing how much headspace you have in a tank and understanding the laws of physics that are associated with the solubility of carbon dioxide in the beer.

Henry's law

Henry's law is partially responsible for the forced carbonation charts that brewer's use. It describes the solubility of a gas (e.g., carbon dioxide) in a solution in relation to the partial pressure of that gas. The higher the pressure, the more CO_2 that is dissolved in solution.

24.3.3 What adjustments did you need to make from working in a lab setting to a brewery setting?

The biggest one was waking up earlier. I did not have to wake up to be at work as early as 6 before. Other than that just being used to there being a sustained need for a product and forecasting that several weeks out. Coming from my previous jobs of working in a coffee shop or as a tour guide the biggest adjustment is that the stakes are much higher. If you mess up a batch of beer it's a lot different than pouring a cup of coffee down the drain.

24.3.4 What do you enjoy most about your current position?

The community at work. We have a solid group of people and I really like working with them day in and day out.

24.3.5 Do you have any words of wisdom for students wishing to apply their chemistry (STEM) degree toward getting into the brewing industry?

I wish I had a good answer for that but I was not a STEM major.

24.4 Experiences in the brewing industry

24.4.1 How would you describe your leadership style?

I would say collaborative within the production area. I am the only person running the lab. In the brewery we are the most collective and collaborative group. When we start talking about production schedules or how to solve a problem it's never quite as one person telling somebody else to do something. It is closer to having a dialog about how to approach a particular problem.

24.4.2 Have you had any negative experiences in the brewing industry that you would like to share?

When I was new and this is advice for anybody trying to get into the industry in general. I took the wrong clamp off of a tank and spilled 1,000 gallons of 29 degree Pilsner all over myself. That woke me up and on top of smelling and having beer all over me, all of the beer was gone.

24.4.3 What are the various ways that you keep yourself grounded and you use to take care of yourself?

I think the biggest rule is just to keep in mind that it is just beer and it is good to just remind yourself of that every once and awhile.

24.4.4 Do you have any advice that you would give to your younger self before getting into the brewing industry?

Yeah, I would say do not work for free. If they tell you to come in and they will try to find you a position avoid doing that. Everyone deserves to be fairly compensated.

24.4.5 Do you have any tips or tricks for maintaining a positive work-life balance?

Do not be afraid to take time off. Just because you can hang out at the brewery and drink free beer it does not mean that you should.

24.4.6 Describe your toughest day in the brewing industry so far

My toughest day was when I was starting out. I was the delivery driver. We decided to combine the Santa Fe and Albuquerque beer delivery runs together. It took fourteen hours to finish those runs.

24.5 Professional affiliations

24.5.1 What current professional organizations are you affiliated with? Why did you join them?

I am a member of ASBC. Specifically to get access to all of their protocols and resources. Also, to learn how to setup and run a good beer lab. In addition to having access to their journal. Some of it may not be immediately applicable to what you are doing or need, but it is still very interesting stuff to read. In addition to that I am a member of the BA. My brewery is also a member of the New Mexico Brewers' Guild. They are a great resource for local networking and they throw all sorts of events all year long.

New Mexico Brewer's Guild
The New Mexico Brewer's Guild started in the 1990s and provides New Mexico brewers with a place to share their passion for craft beer. Their Mission Statement states that "the guild promotes and protects brewers while helping to cultivate a collaborative spirit across and beyond the state (*sic*)."
https://nmbeer.org/about-2
08/09/23

24.5.2 What do you enjoy most about going to their events?

The biggest ones that I enjoyed was the Craft Brewer's Conference in Minneapolis. That was great because it is how the GABF would be put on for breweries. It is put on by suppliers for brewers. You know to look at giant machinery that we may not be buying soon and to demo that and talk to people in the industry about what they are doing.

Craft Brewers Conference
The Craft Brewers Conference is sponsored by the Brewer's Association and is self-described as the "number one environment in North America for concentrated, affordable brewing education." Seminars are offered in a variety of tracks.
https://www.brewersassociation.org/programs/brewers-association-events/craft-brewers-conference/
08/09/23

24.5.3 How important is it to be associated with other professional organizations?

I think it never hurts to network and to get to meet people. The more you know in the industry the better.

24.6 Other/miscellaneous

24.6.1 Outside of brewing what other hobbies or interests do you have?

I have been trying to learn how to fly fish but I have not caught one yet. I have also gotten into making cheese lately. It has a lot of analogs to brewing without feeling like you are at work. I also like hiking on the weekends. We're coming toward the end of the season, but I have been paddle boarding down the Rio Grande a lot.

24.6.2 Is there anything else that you'd like to share about your personal or professional life that you think people might be interested in knowing?

Should you find yourself in a brewery everything will be overwhelming. Never be afraid to ask questions and if somebody starts yelling at you just know that they are wrong and you are not.

25 Oliver Meinhold

MS (Equivalent), Berlin Institute of Technology 1997

Zee Loeffler, Member of The Vincit Group

Berlin Institute of Technology	Zee Loeffler
https://www.tu.berlin/webre	http://www.vincitgroup.com/chemical-application/food-and-beverage/
launch/	brewery-sanitation/
09/02/23	09/02/23

25.1 College/degree pursuit

25.1.1 What inspired you to pursue a chemistry (STEM) degree?

My father was an engineer, and I always liked natural science and wanted to do something that was not that common. I was an exchange student and got more involved in brewing. In the old days it was an Engineering degree (Dipl.-Ing.) in Germany, and now it's the equivalent of a master's degree.

Engineering degree (Dipl.-Ing.)
This is what we know as an Engineering Degree with an MS in the states.

25.1.2 How long did it take you to graduate? Were there any stumbling blocks you'd like to share?

It took me 5½ years. It was a very broad spectrum of courses from organic to inorganic to microbiology to math, physics and process technology. Math can sometimes be difficult to work with. We got more into process and technology. Surprisingly, I thought Math was the most difficult one for me. The main part, of course, was malting and brewing technology.

https://doi.org/10.1515/9783110798777-025

25.1.3 Did you do any research as an undergraduate?

When I finished my master's thesis I did a research topic for that. It was unique because it dealt with returned bottles and looked at the interaction of the bottles with surface acting ingredients in the cleaning products and head retention. In Germany, head retention is very important and has a higher QA focus than over there. When the bottles get cleaned you do not want to waste too much water doing that with rinsing. That was an interesting challenge and so I did various research studies on that and different applications and rinse times and the overall effect on head retention. When I finished that I went to the VLB (Research and Teaching Institute for Brewing in Berlin) which is a famous institution that supported the university. They were noncommercial and were funded out of the industry, and they worked hand in hand with the university. We did some work in that area. Then I switched to the Berlin Institute of technology and was an assistant professor for 7 years.

VLB (Research and Teaching Institute for Brewing in Berlin)
Dedicated to teaching the art and science of brewing since 1883. They focus on
education, research and services for all aspects of the brewing industry.
https://www.vlb-berlin.org/en/home
09/02/23

25.1.4 Do you have any words of wisdom for students wishing to pursue a chemistry (STEM) degree?

What I noticed when I was an assistant professor working with students was that math was very important. You need to be able to calculate percentages and portions correctly. From my daily work that is very important particularly in the brewing industry. You need to be flexible and versatile in many fields. Brewing is a very broad-spectrum field. If you look at beer in Bavaria, it is actually considered food and you can consume it in the cafeteria. Beer is actually one of the most complex foods out there.

Bavarian Beer Is Food!
Oliver was not lying. Beer is literally considered "liquid bread" in Bavaria. Price increases
of this staple have led to rebellions and king abdications in fact!
https://www.nytimes.com/1985/03/03/travel/in-bavaria-beer-is-both-food-and-drink.
html
09/02/23

25.1.5 How often do you use what you learned as an undergraduate in your current job?

Very frequently. In addition to cleaning, we also provide enzymes and brewing supplies. We deal with questions from breweries on how to increase the process aspects of beer in the mash tun and lauter tun with enzyme additions. Also, we help them with how to improve fermentation by maybe reducing the time of the diacetyl rest, so I am more like a manager, but I am also a technical director. A good portion of my work deals with that type of problem solving.

25.2 Academic experience

25.2.1 Could you discuss your work in the brewing science department?

I was more of a manager at that time. I would represent the chair at conferences and do lectures and seminars. Most exams were oral and you are the second person in the room. I had to make a note of the student answers. I answered questions surrounding the lecture. There was not enough time for research projects, but we did look at head retention in a bigger research project. Another major study was on gushing one particular year. The German Brewer's Association provided emergency funds for us to work on that for them. We did joint research on that and the interest waned since the following year gushing was not an issue. Malting was another interest to the bigger beer producers because they were doing batches back to back. Hops was another area that I did some work on there. A major difference between research in Germany and here is that it was more fundamental over there. In the states they want more applied projects to be done.

German Brewer's Association
They like to boast that members generated 3.5 million hectoliters of beer in 1 year.
Members include InBev, Warsteiner and Bitburger.
https://brauer-bund.de/
09/02/23

25.2.2 What classes did you teach?

Everything between malting and brewing. So from the barley coming in to the beer going out in a bottle, can or keg.

25.2.3 Did you enjoy academia? Why or why not?

I loved working in academia and the freedom associated with it. I certainly carried the productivity and troubleshooting lessons over to what I am doing now. We got the green card through the lottery, and moving to the states was always a dream of mine. I like being more able to deal with people and handle management. As a technical director I have to go to a place and help analyze a problem such as foam. You have to go there in person and use all your senses and look for off-pictures. I really like that part of my current job. You have to learn things in person as opposed to from a book.

25.3 Movement to brewing field

25.3.1 What was your "entry beer" into enjoying craft beer? Why did you like it?

I would say that the number one type of beer for me that got me interested in craft beer is an IPA. It brings the dominance of the hop aroma and it makes it so much better. The hop character makes it so good. When you look into the hazy IPAs which is more fine-tuned as opposed to the West Coast ones that are so strong and so intense. Drinkability is very important.

IPA Style Guidelines (BJCP)
There are 10 different subcategories in the 2021 guidelines. ABV values range between 3.0% (Session) to Double (10.0%). This category certainly has grown since I started judging beers.
https://www.bjcp.org/style/2021/21/
09/02/23

25.3.2 Did you ever homebrew?

No I did not,

25.3.3 Prior to getting into brewing what else did you do after graduating?

I actually moved right into it after graduating. This is a big difference between Germany and America. Once you find something you like in Germany you do it for the rest of your life whereas over here people jump around a little. It is hard for people from Germany to understand say somebody working in the banking field, moving into the brewing industry for a while and then moving back into the banking field.

25.4 Current position

25.4.1 What are your current job responsibilities and tasks?

I wear multiple hats and one of those is acting as the technical director. I am in charge of any problem that the sales crew cannot handle, and they ask me to come in. I am also the face of the company, so on big sales they put me in as an additional contact for more complex questions. I represent the company at trade shows and conferences. I attended the MBAA district meeting in Southern California and gave a presentation with a local guy. I teach some courses for MBAA in the maintenance and brewing areas for 7 years now. Recently, I became a regional manager for the West Coast, Wyoming and Montana. I oversee the sales crews there as well.

25.4.2 What are your favorite and least favorite things to do and why?

I love to present and discuss things with customers and brewers. I like to share information and listen to what they are doing. It is good to listen to what challenges and issues they have or if, for example, they have a bad batch of malt we look to maybe fix that with an enzyme. I like interacting with them and learning as well as sharing my experience and lending a helping hand.

25.4.3 What classes that you took apply to some of the things that you have done in the brewing industry? Why/how?

I would say that the malting and brewing courses for sure. I also took some business classes so that helps with discussing money and costs with people. Wastewater treatment is now a big thing, and it is more important than ever with cities asking breweries about phosphates, BODs (Biochemical, or Biological, Oxygen Demand) and the like.

Waste Water Treatment
I know we've already mentioned this in another chapter, but it is worth mentioning again. The BOD value for pre-treated brewery effluent was shown to range between 100 and 400 ppm in this manual.
https://www.brewersassociation.org/attachments/0001/1517/Sustainability_-_Water_Wastewater.pdf
09/02/23

25.4.4 What adjustments did you need to make from working in an academic to a brewery setting?

It was basically driven by life circumstances. When we got our green card I said okay, and now I need to find a job. I got a job in a brewery and that did not quite work out. It might have been a cultural thing. Once I had my foot in the door in the field, my boss at my current company was German as well and hired me. After so many years in academics I was kind of tired. I liked the freedom to do what I wanted, but it was also very exhausting. If you wait for financial support on a project, that can be a problem. Some people can do it for their whole life, but I just did not see myself doing that.

25.4.5 What do you enjoy most about your current position?

I like to see different breweries, different people and how they approach things differently. Some are more regimented and focused on quality and some are more creative. I really like those differences and a nice picture of the entire industry.

25.4.6 Do you have any words of wisdom for students wishing to apply their chemistry (STEM) degree toward getting into the brewing industry?

Definitely focus on the science and the biochemistry. The brewing steps are so tied into those. You need to have the will to work hard. There is quite a bit of manual labor that goes into it. As more stuff gets automated it has gotten a little easier. I remember when you had to work with 50 kg bags and now it is down to 25 kg bags of malt. When I was an intern I had to go into tanks and OSHA would not like that. Keep in mind that 80% of the work in the brewing industry is cleaning and sanitation. They need to understand what is going on in the industry. In Germany you actually do an internship so you know what to do and what it is like.

Occupational Safety and Health Administration (OSHA)
Regardless of your thoughts on this four letter word to some, they do provide some
excellent resources to help keep your workplace safe.
https://www.osha.gov/
09/02/23

25.5 Experiences in the brewing industry

25.5.1 How would you describe your leadership style?

I am more of the supportive side or style of leader. Sometimes you have to give an order of course. Mostly though, I will try to guide my team and give them a chance to fix something or address it. I might ask them to give a presentation and then give them hints and advice. It is more like less authoritarian and more open. Guidelines and deadlines are important though, but whenever there is a problem I like to give them a chance to learn from it. It is good for my colleagues to walk a customer through a problem and actually learn from that process themselves.

25.5.2 What are the various ways that you keep yourself grounded and you use to take care of yourself?

I think it's my natural personality to never go crazy or exceed things. When you get asked back then it is a good thing. I try not to make up things. When I'm asked a question I cannot answer then I tell them that I will look into it. Honesty is very important. The craft industry has so many people willing to share and discuss things- just look at all the collaboration beers. Interestingly, the one area where people are not as open right now though is the nonalcoholic beers. If you properly represent yourself and they invite you for training and they ask you to come back then that really helps.

25.5.3 Have you had any negative experiences in the brewing industry that you would like to share?

A negative is always when something does not happen like you want it. Like you do not get a big deal or lose a customer. Then you need to analyze what happened and try to understand it. When you do not get a deal after 3 years from a big customer that is a sad thing and negative. Other than that I think it is best to be about customers and work hard to get and keep a customer.

25.5.4 Do you have any advice that you would give to your younger self before getting into the brewing industry?

For people who are thinking about what to do I would recommend working in the brewing industry to get a better picture of how everything works. People think it is romantic and they have a very idealistic view of it. Like I said, most of it is cleaning and sanitation. There is a lot of hard labor as opposed to just pushing buttons. You need to have a clear picture and can get that from an internship in the industry. I had a lot of people who did this and ended up loving it. They then took a chance to go back to school and climbed up the ladder.

25.6 Professional affiliations

25.6.1 What current professional organizations are you affiliated with? Why did you join them?

I was a recent member of ASBC. I'm a member of MBAA and as a technical director I am required to do that. BA is a company membership that pays for everyone so I don't need an individual membership. When you do a trade show at say the Craft Brewers Conference what is so neat is that you learn and see new things. You get to listen to the latest research. I like getting those updates from a professional benefit as well. I like staying on top of things and staying in touch with people. What I learned is that you need to be present on a regular basis, but if you stop showing up then people notice and they may start saying negative things. It is important to be there and to network. I am still a member of the German Brewing and Maltster Association. They are like the MBAA in that they are a national organization and they meet annually.

German Brewing and Maltster Association
This was a hard group to find much information on. The referenced article describes a meeting where the group hosted a party at m + f KEG-Technik to celebrate their 50th anniversary.
https://www.brewersjournal.info/german-brew-masters-and-maltsters-association-dbmb-gather-in-bottrop/
09/02/23

25.6.2 What do you enjoy most about going to their events?

I like sharing and exchanging those experiences as well as my knowledge about things. I noticed that you are often asked to give equipment recommendations. I work with breweries on their plans and take a look at where I can help with them and we share the common interest of brewing as well as networking.

25.6.3 How important is it to be associated with other professional organizations?

I think if you want to join the brewing related ones it is a good way to get your name out there. For younger ones we tell them to be an MBAA member and be active in their district. This is where you can make yourself a good name in the industry.

25.7 Other/miscellaneous

25.7.1 Outside of brewing what other hobbies or interests do you have?

I have a young daughter, and now that she is older I can go and golf some more. Interestingly, Boston Brewing Company asked me to do a golf tournament with them. I have a friend working there, and now I am actually playing golf with them. I also like to travel and I can combine that with work. When it is possible I take my wife and daughter with me. They can explore the city where the meeting is. I also like to go back to Germany since we have a couple of companies over there. I like the outdoors and a good part of the job allows me to combine my work with wining and dining. I love to eat and going out is fun. We also like to travel to go meet people as well.

25.7.2 Is there anything else that you'd like to share about your personal or professional life that you think people might be interested in knowing?

What I was told several times is that since I am from Germany and still have the accent, this is a benefit to me. I have been told that it gives me more trust and credibility. It can act as a door opener. It is not that I use it for marketing but it is definitely a big plus. Beer in Germany is older and so much equipment still comes from there. Famous suppliers are from there as well so it is something that everyone is familiar with. It certainly does help and makes my work a little easier.

Index

https://doi.org/10.1515/9783110798777-026

www.ingramcontent.com/pod-product-compliance
Lightning Source LLC
Chambersburg PA
CBHW080930220326
41598CB00034B/5745